Elementary Linear Algebra
Second Edition

Elementary Linear Algebra

Second Edition

Devi Prasad

Alpha Science International
Oxford, U.K.

Elementary Linear Algebra
Second Edition
198 pgs.

Devi Prasad
Professor
Department of Computer Science and Engineering
Krishna Engineering College
Ghaziabad

Copyright © 2006, 2012
Second Edition 2012

ALPHA SCIENCE INTERNATIONAL LTD.
7200 The Quorum, Oxford Business Park North
Garsington Road, Oxford OX4 2JZ, U.K.

www.alphasci.com

All rights reserved. No part of this publication may be reproduced, stored in a retrieval system, or transmitted in any form or by any means, electronic, mechanical, photocopying, recording or otherwise, without prior written permission of the publisher.

ISBN 978-1-84265-400-2

Printed in India

To My Wife Chandra Vati

Preface to the Second Edition

The book is well-organized, lucidly written text introduces the reader to **Elementary Linear Algebra.** Suggestions to add more solved examples and addition of matrix method to check the linear dependence and independence came from several readers and my colleagues. The above suggestions have been incorporated in this second edition and contents have been rearranged for smooth understanding and better readability of the subject. A large number of illustrated examples are covered to clarify the theoretical concepts and many unsolved problems are also given for practice to enhance the presentation of the material. The book is suitable for undergraduate students of science and engineering. It gives me immense pleasure to convey that the first edition of the book was very much liked by the students as they could understand the subject easily due to smooth flow of various topics in the book.

Suggestions for improving the contents will be highly appreciated.

Devi Prasad

Preface to the First Edition

This book 'Elementary Linear Algebra' is the outcome of my experience of teaching Linear Algebra at undergraduate level for several years in the Birla Institute of Technology & Science, Pilani. In view of this the presentation of the subject matter is smooth and easily understandable by the reader. Proof of few theorems are presented. The book is suitable for students of Engineering & Science disciplines.

The book consists of solving system of linear equations in the first chapter, vector spaces in second chapter, linear transformations in third chapter, eigenvalues and eigenvectors in fourth chapter and finally inner product in fifth chapter. At the end of each chapter sufficient unsolved problems are given for the practice of students.

I convey my sincere thanks to Prof. S Venkateswaran Vice-Chancellor of the institute for giving me an opportunity of writing the text book. My sincere thanks are due to Prof. L K Maheshwari Pro-Vice-Chancellor & Director of the institute for encouraging me to write the book. My thanks are due to Prof. A K Sarkar for providing me the infrastructural facilities for preparing the manuscript of the book.

I extend my heartiest thanks to Dr Nadeem-Ur-Rehman and Dr A Vasan for their suggestions.

Finally, I thank Shri Dalbag Singh, who word processed the book with patience and care.

Devi Prasad

Contents

Preface to the Second Edition   vii
Preface to the First Edition   ix

1. Matrices and Algebraic Structure   1.1
   1.1 Definitions   1.1
   1.2 Addition and Multiplication of Matrices   1.3
   1.3 Transpose of a Matrix   1.5
   1.4 Matrices of Complex Numbers   1.6
   1.5 Groups   1.10
   1.6 Rings   1.20
   1.7 Field   1.23
   Exercise Set 1   *1.25*

2. System of Linear Equations   2.1
   2.1 Solution by Graphs and Elementary Row Operations   2.2
   2.2 Row Reduced Echelon Form   2.9
   2.3 Consistency of the Linear System   2.12
   2.4 Non-singular Matrices and Determinant Values   2.16
   2.5 Inverse of a Matrix   2.20
   Exercise Set 2   *2.28*

3. Vector Spaces   3.1
   3.1 Vector Spaces   3.1
   3.2 Subspace   3.9

3.3 Linear Dependence and Independence ... 3.13
3.4 Basis and Dimension ... 3.26
Exercise Set 3 ... *3.35*

4. Linear Transformations ... **4.1**

4.1 Linear Transformations ... 4.2
4.2 Null and Range Spaces ... 4.7
4.3 Inverse Linear Transformations ... 4.18
4.4 More about Linear Transformations ... 4.23
4.5 Matrices Related to Linear Transformation ... 4.25
4.6 Rank of Matrix ... 4.31
Exercise Set 4 ... *4.33*

5. Eigenvalues and Eigenvectors ... **5.1**

5.1 Eigenvalues and Eigenvectors ... 5.1
5.2 Gershgorin Circle Theorem ... 5.9
5.3 Diagonalization of a Matrix ... 5.14
5.4 Diagonalization of Symmetric Matrices ... 5.17
5.5 Quadratic Forms ... 5.20
Exercise Set 5 ... *5.25*

6. Inner Product ... **6.1**

6.1 Inner Product ... 6.1
6.2 Orthogonality ... 6.3
6.3 Gram-Schmidt Orthgonalization ... 6.5
6.4 Inner Product Spaces ... 6.11
Exercise Set 6 ... *6.12*

Index ... **I.1**

CHAPTER 1

Matrices and Algebraic Structure

1.1 DEFINITIONS

Matrix: A matrix is an arrangement of $m \times n$ numbers in a rectangular form within a square or small brackets as shown below where a_{ij} are some numbers:

$$\begin{bmatrix} a_{11} & a_{12} & \cdots & a_{1n} \\ a_{21} & a_{22} & \cdots & a_{2n} \\ \vdots & \vdots & \vdots & \vdots \\ a_{i1} & a_{i2} & \cdots & a_{in} \\ \vdots & \vdots & \vdots & \vdots \\ a_{m1} & a_{m2} & \cdots & a_{mm} \end{bmatrix}$$ and is also denoted by $A = (aij)_{m \times n}$ short notation.

First suffix of a_{ij} denotes row and second suffix denotes column.

Each number a_{ij} is called element or entry of the matrix. Horizontal entries are called rows and vertical entries are called columns. In above matrix number of rows is m and number of columns is n and is known m by n matrix. Matrix does not have any value, this is only an arrangement of numbers until we associate this with some other mathematical expressions.

Entries a_{ii} $j = i$ are known diagonal and a_{ij}, $i \neq j$ are known as off-diagonal.

Example 1: $A = \begin{bmatrix} 3 & 1 & -5 & 7 \\ 2 & -5 & 3 & 8 \\ -4 & 2 & 6 & 7 \end{bmatrix}$ is a matrix with 3 rows and 4 columns.

Multiplication of a matrix by a number.

If a matrix is multiplied by a constant then each entry is multiplied by that constant.

From example above $4A = \begin{bmatrix} 12 & 4 & -20 & 28 \\ 8 & -20 & 12 & 32 \\ -16 & 8 & 24 & 28 \end{bmatrix}$, $-A = \begin{bmatrix} -3 & -1 & 5 & -7 \\ -2 & 5 & -3 & -8 \\ 4 & -2 & -6 & -7 \end{bmatrix}$

Similarly for a given matrix B

$B = \begin{bmatrix} 2 & 4 & -3 \\ 5 & -4 & 7 \\ 6 & 1 & 3 \\ -1 & 3 & 2 \end{bmatrix}$ is 4×3 matrix, $-2B = \begin{bmatrix} -4 & -8 & 6 \\ -10 & 8 & -14 \\ -12 & -2 & -6 \\ 2 & -6 & -4 \end{bmatrix}$

Square Matrix: If the number of rows is equal to the number of columns i.e., $m = n$, then the matrix is known as square matrix.

Diagonal Matrix: A square matrix of the form $\begin{bmatrix} a_{11} & 0 & \cdots & 0 \\ 0 & a_{22} & \cdots & 0 \\ \vdots & \vdots & \vdots & \vdots \\ 0 & 0 & \cdots & 0 \\ \vdots & \vdots & \vdots & \vdots \\ 0 & 0 & \cdots & a_{nn} \end{bmatrix}$

with $a_{ij} = 0$ for $i \neq j$ and at least one of the diagonal entry nonzero is called diagonal matrix.

The matrix $\begin{bmatrix} 12 & 0 & 0 & 0 \\ 0 & -1 & 0 & 0 \\ 0 & 0 & 5 & 0 \\ 0 & 0 & 0 & 8 \end{bmatrix}$ is a diagonal matrix.

If all the entries of a matrix are zeros then it is known zero matrix. $\begin{bmatrix} 0 & 0 & 0 & 0 \\ 0 & 0 & 0 & 0 \\ 0 & 0 & 0 & 0 \\ 0 & 0 & 0 & 0 \end{bmatrix}$

Identity Matrix or unit matrix: A diagonal matrix with $a_{ii} = 1$, for all $i = 1,2,3...n$ is called Identity Matrix or unit matrix.

Example 2. $A = \begin{bmatrix} 1 & 0 & 0 & 0 \\ 0 & 1 & 0 & 0 \\ 0 & 0 & 1 & 0 \\ 0 & 0 & 0 & 1 \end{bmatrix}$ is unit matrix of order 4.

A square matrix $\begin{bmatrix} a_{11} & a_{12} & \cdots & a_{1n} \\ 0 & a_{22} & \cdots & a_{2n} \\ \vdots & \vdots & \vdots & \vdots \\ 0 & 0 & \cdots & a_{in} \\ \vdots & \vdots & \vdots & \vdots \\ 0 & 0 & \cdots & a_{nn} \end{bmatrix}$ with $a_{ij} = 0$ for $i > j$ is called upper triangular matrix

i.e. all entries below diagonal entries are zeros and similarly lower triangular if $a_{ij} = 0$ for $i < j$ i.e., all entries above diagonal entries are zeros.

Example 3. $\begin{bmatrix} 3 & 5 & 6 & 7 \\ 0 & -4 & 3 & 0 \\ 0 & 0 & 0 & 2 \\ 0 & 0 & 0 & 5 \end{bmatrix}$ is upper triangular since all entries below diagonal are zeros

and $\begin{bmatrix} 3 & 0 & 0 & 0 \\ 5 & -4 & 0 & 0 \\ 6 & 0 & 0 & 0 \\ -4 & 12 & 5 & -2 \end{bmatrix}$ is lower triangular since all the entries above diagonal are zeros.

Upper or lower triangular Matrices with all diagonal entries are called strictly upper or strictly lower triangular Matrices.

1.2 ADDITION AND MULTIPLICATION OF MATRICES

Addition: Two matrices can be added if they are of the same order i.e., of same size. Number of rows in both the matrices are same and number of columns of both are also same. Let $A = (a_{ij})_{m \times n}$ and $B = (b_{ij})_{m \times n}$ be two matrices of the same order.

Addition or subtraction of these two is as $A \pm B = (a_{ij} \pm b_{ij})_{m \times n}$, in short add corresponding entries of the two matrices.

Example 4: $\begin{bmatrix} 2 & 4 & -4 \\ 5 & -3 & 6 \\ -4 & 5 & 3 \end{bmatrix} + \begin{bmatrix} 3 & -2 & 5 \\ -3 & 2 & 1 \\ 7 & 2 & 4 \end{bmatrix} = \begin{bmatrix} 5 & 2 & 1 \\ 2 & -1 & 7 \\ 3 & 7 & 7 \end{bmatrix}.$

Multiplication of matrices: Two matrices can be multiplied if the number of columns of first matrix is equal to the number of rows of second matrix. If matrix $A_{m \times n}$ has n columns and $B_{n \times p}$ has n rows, then matrix multiplication AB is defined, yet BA may not be defined.

Let $A = \begin{bmatrix} a_{11} & a_{12} & \cdots & a_{1n} \\ a_{21} & a_{22} & \cdots & a_{2n} \\ \vdots & \vdots & \vdots & \vdots \\ a_{i1} & a_{i2} & \cdots & a_{in} \\ \vdots & \vdots & \vdots & \vdots \\ a_{m1} & a_{m2} & \cdots & a_{mn} \end{bmatrix}$, $B = \begin{bmatrix} b_{11} & b_{12} & \cdots & a_{1p} \\ b_{21} & b_{22} & \cdots & b_{2p} \\ \vdots & \vdots & \vdots & \vdots \\ b_{j1} & b_{j2} & \cdots & b_{jp} \\ \vdots & \vdots & \vdots & \vdots \\ b_{n1} & b_{n2} & \cdots & n_{np} \end{bmatrix}$ be two matrices of order $m \times n$

and $n \times p$ respectively. Further let $C = AB$, where $C = (c_{ik})_{m \times p}$.

Now to get the entry of ith row and k^{th} column of the matrix C, pick up entries of ith row of matrix A and k^{th} column of matrix B, then multiply the corresponding entries and add, then

$c_{ik} = a_{i1}b_{1k} + a_{i2}b_{2k} + \ldots + a_{in-1}b_{n-1k} + a_{in}b_{nk}$, for $i = 1,2,3\ldots, (m-2), (m-1), m$ and $k = 1,2,3\ldots, (p-2), (p-1), p$. Matrix C will have m rows and p columns.

Example 5: Let $A = \begin{bmatrix} 2 & 1 & -3 & 4 \\ 3 & 2 & -1 & 2 \\ 1 & -3 & 4 & 2 \end{bmatrix}$ and $B = \begin{bmatrix} 3 & -2 \\ 4 & 2 \\ -1 & 3 \\ 2 & 1 \end{bmatrix}$. Then compute AB.

Solution: $AB = \begin{bmatrix} 2 & 1 & -3 & 4 \\ 3 & 2 & -1 & 2 \\ 1 & -3 & 4 & 2 \end{bmatrix} \begin{bmatrix} 3 & -2 \\ 4 & 2 \\ -1 & 3 \\ 2 & 1 \end{bmatrix}$

$= \begin{bmatrix} 6+4+3+8 & -4+2-9+4 \\ 9+8+1+4 & -6+4-3+2 \\ 3-12-4+4 & -2-6+12+2 \end{bmatrix} = \begin{bmatrix} 21 & -7 \\ 22 & -3 \\ -9 & 6 \end{bmatrix}$

Note: This is 3 by 2 matrix.

For given matrices A, B if matrix multiplication AB is defined, then BA may not be defined. In the above example BA is not defined due to unequal number of columns of B and number of rows of A.

In case of square matrices AB and BA both are defined but in general $AB \neq BA$.

1.3 TRANSPOSE OF A MATRIX

Transpose: A matrix $B = (b_{ij})$ is called transpose of matrix $A = (a_{ij})$, if $b_{ij} = a_{ji}$ for all i,j. Rows of B are columns of A. Transpose of A is denoted by A^T or A'.

Example 6: Let $A = \begin{bmatrix} 2 & -3 & 0 & 5 \\ 4 & 6 & -2 & 1 \\ 5 & 2 & 3 & 4 \end{bmatrix}$, then $A' = \begin{bmatrix} 2 & 4 & 5 \\ -3 & 6 & 2 \\ 0 & -2 & 3 \\ 5 & 1 & 4 \end{bmatrix}$.

It is easy to see that $(A \pm B)' = A' \pm B'$ and also $(A')' = A$.

Symmetric Matrix: A square matrix $A_{n \times n}$ is known symmetric if $A'_{n \times n} = A$ i.e. $a_{ij} = a_{ji}$ for all, $i,j = 1,2,3...n$.

Example 7: Let $A = \begin{bmatrix} 3 & 1 & -2 & 4 \\ 1 & 2 & 3 & 6 \\ -2 & 3 & -4 & -3 \\ 4 & 6 & -3 & 5 \end{bmatrix}$, then $A' = \begin{bmatrix} 3 & 1 & -2 & 4 \\ 1 & 2 & 3 & 6 \\ -2 & 3 & -4 & -3 \\ 4 & 6 & -3 & 5 \end{bmatrix}$, so A is a symmetric matrix.

Skew-Symmetric Matrix: A square matrix A is known as skew symmetric if $A' = -A$ i.e., For off-diagonal entries $a_{ij} = -a_{ji}$ and diagonal entries are zeros.

Example 8: Let $A = \begin{bmatrix} 0 & -1 & 2 & -4 \\ 1 & 0 & -3 & -6 \\ -2 & 3 & 0 & 3 \\ 4 & 6 & -3 & 0 \end{bmatrix}$, then $A' = \begin{bmatrix} 0 & 1 & -2 & 4 \\ -1 & 0 & 3 & 6 \\ 2 & -3 & 0 & -3 \\ -4 & -6 & 3 & 0 \end{bmatrix}$,

$-A = \begin{bmatrix} 0 & 1 & -2 & 4 \\ -1 & 0 & 3 & 6 \\ 2 & -3 & 0 & -3 \\ -4 & -6 & 3 & 0 \end{bmatrix} = A'$, therefore A is skew–symmetric.

1.6 *Elementary Linear Algebra*

Let $A = (a_{ij})_{m \times n}$, $B = (b_{jk})_{n \times p}$ then $AB = (c_{ik})_{m \times p}$. $A'_{n \times m}$ is n by m and $B'_{P \times N}$ is p by n, therefore multiplication $B'A'$ is defined and $(AB)' = B'A'$, this can be shown by computing both sides, involve simple work but too much.

Every matrix A can be written as sum of symmetric and skew–symmetric matrices.

We can write $A = \frac{1}{2}(A + A') + \frac{1}{2}(A - A')$.

Now $\frac{1}{2}(A + A')' = \frac{1}{2}(A' + (A')') = \frac{1}{2}(A' + A) = \frac{1}{2}(A + A')$, so this part is symmetric.

Similarly $\frac{1}{2}(A - A')' = \frac{1}{2}(A' - (A')') = \frac{1}{2}(A' - A) = -\frac{1}{2}(A - A')$ this part is skew symmetric.

1.4 MATRICES OF COMPLEX NUMBERS

Like matrices of real number we can also have matrices of complex numbers, where entries are complex numbers.

Example 9: $A = \begin{bmatrix} 2 - 3i & 3 + i & 1 - 4i \\ 3 + 2i & 4 - 3i & 2 + i \\ 6 & 5i & 1 - i \end{bmatrix}$ is matrix of complex numbers.

If the entries of the matrix A are complex numbers, matrix obtained by taking complex conjugate of the numbers of the given matrix is denoted by $\overline{A}$.

Complex conjugate of above matrix is $\overline{A} = \begin{bmatrix} 2 + 3i & 3 - i & 1 + 4i \\ 3 - 2i & 4 + 3i & 2 - i \\ 6 & -5i & 1 + i \end{bmatrix}$

Transpose of complex conjugate $(\overline{A})'$ = conjugate of (A') is denoted by A^θ, some other notation are also used in some other books.

Hermitian Matrix: A matrix A is known Hermitian, if $A^\theta = A$.

Example 10: Let $A = \begin{bmatrix} 3 & 2 - 4i & 4 + 2i & 5 - 3i \\ 2 + 4i & -4 & 6 - i & 3 + 2i \\ 4 - 2i & 6 + i & 2 & 1 - 5i \\ 5 + 3i & 3 - 2i & 1 + 5i & 5 \end{bmatrix}$. Show that A is Hermitian.

Solution:

$$\overline{A} = \begin{bmatrix} 3 & 2+4i & 4-2i & 5+3i \\ 2-4i & -4 & 6+i & 3-2i \\ 4+2i & 6-i & 2 & 1+5i \\ 5-3i & 3+2i & 1-5i & 5 \end{bmatrix}$$

So $(\overline{A})' = \begin{bmatrix} 3 & 2-4i & 4+2i & 5-3i \\ 2+4i & -4 & 6-i & 3+2i \\ 4-2i & 6+i & 2 & 1-5i \\ 5+3i & 3-2i & 1+5i & 5 \end{bmatrix} = A$, A is Hermitian.

Skew-Hermitian Matrix: A matrix A is known Skew-Hermitian, if $A^\theta = -A$.

Example 11: If $A = \begin{bmatrix} 2+3i & -1+3i \\ -5i & 4-2i \end{bmatrix}$, show that $A^\theta A$ is Hermitian matrix

Solution: $\overline{A} = \begin{bmatrix} 2-3i & -1-3i \\ 5i & 4+2i \end{bmatrix}$, $\overline{A}' = \begin{bmatrix} 2-3i & 5i \\ -1-3i & 4+2i \end{bmatrix}$,

so $A^\theta = \begin{bmatrix} 2-3i & 5i \\ -1-3i & 4+2i \end{bmatrix}$.

$A^\theta A = \begin{bmatrix} 2-3i & 5i \\ -1-3i & 4+2i \end{bmatrix} \begin{bmatrix} 2+3i & -1+3i \\ -5i & 4-2i \end{bmatrix} =$

$\begin{bmatrix} (2-3i)(2+3i)+25 & (2-3i)(-1+3i)+5i(4-2i) \\ (-1-3i)(2+3i)+(4+2i)(-5i) & (-1-3i)(-1+3i)+(4+2i)(4-2i) \end{bmatrix}$

$= \begin{bmatrix} 38 & 17+29i \\ 17-29i & 30 \end{bmatrix} = B$ say

$\overline{B} = \begin{bmatrix} 38 & 17-29i \\ 17+29i & 30 \end{bmatrix}$, $\overline{B}' = \begin{bmatrix} 38 & 17+29i \\ 17-29i & 30 \end{bmatrix}$, $B^\theta = \begin{bmatrix} 38 & 17+29i \\ 17-29i & 30 \end{bmatrix} = B$.

Therefore $B = A^\theta A$ is Hermitian.

Example 12: If $A = \begin{bmatrix} 2+i & 3 & -1+3i \\ -5 & 1 & 4-2i \end{bmatrix}$, show that $A^\theta A$ is a Hermitian matrix.

Solution: $\overline{A} = \begin{bmatrix} 2-i & 3 & -1-3i \\ -5 & 1 & 4+2i \end{bmatrix}$ $\overline{A}' = \begin{bmatrix} 2-i & -5 \\ 3 & 1 \\ -1-3i & 4+2i \end{bmatrix}$

$$\overline{A}'A = \begin{bmatrix} 2-i & -5 \\ 3 & 1 \\ -1-3i & 4+2i \end{bmatrix} \begin{bmatrix} 2+i & 3 & -1+3i \\ -5 & 1 & 4-2i \end{bmatrix}$$

$$= \begin{bmatrix} 10 & 1-3i & -19+17i \\ 1+3i & 5 & 5+7i \\ -19-17i & 5-7i & 30 \end{bmatrix} = B(\text{say})$$

Let $B = \overline{A}'A$. $\overline{B} = \begin{bmatrix} 10 & 1+3i & -19-17i \\ 1-3i & 5 & 5-7i \\ -19+17i & 5+7i & 30 \end{bmatrix}$

$\overline{B}' = \begin{bmatrix} 10 & 1-3i & -19+17i \\ 1+3i & 5 & 5+7i \\ -19-17i & 5-7i & 30 \end{bmatrix} = B$. Hence $\overline{A}'A$ is Hermitian

Unitary matrix: A matrix A is called Unitary if $A^\theta A = \overline{A}'A = I$

Example 13: Let $A = \begin{bmatrix} 2i & 2-4i & -4+2i & 5-3i \\ -2-4i & 0 & 6-i & 3+2i \\ 4+2i & -6-i & -3i & -1+5i \\ -5-3i & -3+2i & 1+5i & 4i \end{bmatrix}$.

Show that A is Skew-Hermitian.

Solution: $\overline{A} = \begin{bmatrix} -2i & 2+4i & -4-2i & 5+3i \\ -2+4i & 0 & -6+i & 3-2i \\ 4-2i & -6+i & 3i & -1-5i \\ -5+3i & -3-2i & -1-5i & -4i \end{bmatrix}$

$$\text{So } \overline{(A)}' = \begin{bmatrix} -2i & -2+4i & -4-2i & -5+3i \\ 2+4i & 0 & -6+i & -3-2i \\ -4-2i & -6+i & 3i & -1-5i \\ 5+3i & 3-2i & -1-5i & -4i \end{bmatrix}$$

$$= -\begin{bmatrix} 2i & 2-4i & 4+2i & 5-3i \\ -2-4i & 0 & 6-i & 3+2i \\ 4+2i & 6-i & -3i & 1+5i \\ -5-3i & -3+2i & 1+5i & 4i \end{bmatrix} = -A, \text{ so } A \text{ Skew-Hermitian}$$

Example 14: Define Unitary matrix. Show that the matrix $\begin{bmatrix} \alpha+i\gamma & -\beta+i\delta \\ \beta+i\delta & \alpha-i\gamma \end{bmatrix}$ is Unitary matrix, if $\alpha^2 + \beta^2 + \gamma^2 + \delta^2 = 1$

Solution: A matrix A is called Unitary if $A^\theta A = \overline{A}'A = 1$

For the given matrix $A = \begin{bmatrix} \alpha+i\gamma & -\beta+i\delta \\ \beta+i\delta & \alpha-i\gamma \end{bmatrix}$

$$\overline{A} = \begin{bmatrix} \alpha-i\gamma & -\beta-i\delta \\ \beta-i\delta & \alpha+i\gamma \end{bmatrix}$$

$$\overline{A}' = \begin{bmatrix} \alpha-i\gamma & \beta-i\delta \\ -\beta-i\delta & \alpha+i\gamma \end{bmatrix}$$

$$\overline{A}'A = \begin{bmatrix} \alpha-i\gamma & \beta-i\delta \\ -\beta-i\delta & \alpha+i\gamma \end{bmatrix}\begin{bmatrix} \alpha+i\gamma & -\beta+i\delta \\ \beta+i\delta & \alpha-i\gamma \end{bmatrix}$$

$$= \begin{bmatrix} \alpha^2+\gamma^2+\beta^2+\delta^2 & 0 \\ 0 & \beta^2+\delta^2+\alpha^2+\gamma^2 \end{bmatrix}$$

$$= \begin{bmatrix} 1 & 0 \\ 0 & 1 \end{bmatrix}, \text{ if } \alpha^2+\gamma^2+\beta^2+\delta^2 = 1$$

Hence $A = \begin{bmatrix} \alpha+i\gamma & -\beta+i\delta \\ \beta+i\delta & \alpha-i\gamma \end{bmatrix}$ is Unitary.

Every complex matrix A can be written as sum of Hermitian and skew–Hermitian matrices.

We can write $A = \frac{1}{2}(A + A^\theta) + \frac{1}{2}(A - A^\theta)$.

Now $\frac{1}{2}(A + A^\theta)^\theta = \frac{1}{2}(A^\theta + (A^\theta)^\theta) = \frac{1}{2}(A^\theta + A) = \frac{1}{2}(A + A^\theta)$, so this part is Hermitian.

Determinant value of a $n \times n$ matrix.

Let $A = (a_{ij})$ $n \times n$ matrix. Determinant value of A denoted by det $(A) = |A|$ is defined as:
$$\det(A) = \Sigma \pm a_{1j_i} a_{2j_2} \ldots, a_n J_n,$$
where $J_1, J_2 \ldots J_n$ is a permutation of $(1, 2, 3, \ldots, n-1, n)$.

Plus sign is taken if permutation $J_1, J_2 \ldots, J_n$ is even, minus sign is taken if permutation is odd.

Sum is taken over all permutations of $J_1, J_2, \ldots J_n$.

1.5 GROUPS

It is assumed that the reader is familiar with set of real numbers

The following notations will be used very often.

$N = \{1, 2, 3, 4, \ldots\ldots\ldots\ldots\}$ Set of all natural numbers

$Z = \{\ldots\ldots, -4, -3, -2, -1, 01, 2, 3, 4, \ldots \ldots\ldots\ldots\}$. Set of all integers

$Q = \left\{ x = \frac{p}{q}, \text{ where } p,q \text{ are any integers } q \neq 0 \right\}$ Set of all rational numbers.

$R =$ Set of all real numbers

$C =$ Set of all complex numbers

Binary Operation: Let S be a non empty set. An operation denoted by say $*$ is said to be defined on S if for some a and b in S, $a * b$ is an element of same nature as elements of S. If $a * b \in S$ for all $a, b \in S$, then S is closed under operation $*$ or operation $*$ is closed.

Example 15: Let $N = \{1, 2, 3, 4, \ldots\ldots\ldots\ldots\}$ be the set of all natural numbers. Usual addition is an binary operation. Subtraction is not a binary operation, because it may become negative.

Example 16: Let R be the set of all real numbers. Multiplication of any two real numbers is again a real number therefore usual multiplication is an binary operation.

Let S be set and a binary operation say $*$ be defined on S.

(a) Associative law. A Binary operation $*$ is called associative
 if $(a * b) * c = (a*(b*c))$ for all $a, b, c \in S$ 1.1

(b) **Existence of identity element.** If there is an element $e \in S$ such that $a * e = e * a = a$ for all $a \in S$. Then e is called an identity element of set S. 1.2

(c) **Existence of an inverse.** If for each $a \in S$, there is an element x in S such that $a * x = x * a = e$, then x is called inverse of a. x is different for different a. 1.3

(d) Binary operation $*$ is called commutative if $a * b = b * a$ for all $a, b \in S$. 1.4

Semi-group: Let G be a set and a binary operation $*$ be defined on it. If binary operation $*$ is associative (1.1) on G, i.e. $a * (b * c) = (a * b) * (c)$ for all $a, b, c, \in G$. Then $(G, *)$ is called semi-group

Example 17: Let $N = \{1, 2, 3, 4, \ldots \ldots \ldots \}$ be the set of all natural numbers. Usual addition $+$ is an binary operation. Addition of any two natural numbers is again a natural number in N and $+$ is associative. Therefore $(N, +)$ is a semi-group.

Example 18: Let R be the set of all real numbers with usual multiplication. then $(R, .)$ is a semi-group.

Monoid: Let $(G, *)$ be a semi-group. If G has identity element in it, then G is known as monoid that is If there is an element $e \in G$ such that $a * e = e * a = a$, for all $a \in G$, then G is called a monoid. And e is known as identity element of G. In other words if a binary operation on set G satisfies conditions (1.1) and (1.2), then $(G, *)$ is monoid.

Example 19: Let $N = \{1, 2, 3, 4, \ldots \ldots \ldots \}$ be the set of all natural numbers. Usual addition $+$ is an binary operation. Addition of any two natural numbers is again a natural number in N and $+$ is associative. Element 0 is an identity element of addition but not in the set. Therefore $(N, +)$ is not a monoid. If 0 is included in the set then it is monoid.

Example 20: Let R be the set of all real numbers with usual multiplication. Number 1 is an multiplicative identity, therefore $(R, .)$ is a semi-group.

Group: A monoid satisfying condition (1.3) is known as a group. Restated as: if $(G, *)$ set G with binary operation $*$ defined on it, satisfies conditions (1.1), (1.2) and (1.3), then it is known a group.

Example 21: Let R be the set of all real numbers. Usual addition $+$ is an binary operation. Addition of any two real numbers is again a real number in R and $+$ is associative. Element 0 is an identity element of addition. Further every real number has its negative as its additive inverse. Therefore $(R, +)$ is a group.

Example 22: Let R be the set of all real numbers with usual multiplication. Number 1 is an multiplicative identity, but 0 does not have multiplicative inverse. therefore $(R, .)$ is a not group. Set $\{R - 0\}$ with multiplication is a group.

Example 23: Let $[R - 1]$ be the set of all real numbers minus 1, with binary operation $*$ defined by

$a * b = a + b - ab$. Obviously operation is closed.

(i) $(a * b) * c = (a + b - ab) * c = a + b - ab + c - (a + b - ab) \cdot c$
$= a + b + c - ab - ac - bc + abc$

$$a * (b * c) = a * (b + c - bc) = a + b + c - bc - a(b + c - bc)$$
$$= a + b + c - bc - ab - ac + abc = a + b + c - ab - ac - bc + abc.$$

This shows that $(a * b) * c = a * (b * c)$. therefore binary operation is associative.

(ii) 0 is identity element because $a * 0 = a + 0 - 0a = a = 0 * a = 0 + a - 0a$.

(iii) $a + b - ab = 0$, $b(1 - a) = -a$ implies $b = \dfrac{a}{a-1}$. This shows that inverse of a is $\dfrac{a}{a-1}$ when $a \neq 1$.

Therefore all the conditions of the group are satisfied. [R − 1] be the set of all real numbers minus 1, with binary operation * defined by $a * b = a + b - ab$ is a group.

Abelian Group: Let $(G,*)$ be a group. If * operation is commutative, then $(G,*)$ is called abelian group.

Examples 24: Examples 21, 22, 23 are abelian group.i.e., the set of all real numbers with usual sual addition + as an binary operation, Let [R − O] be the set of all real numbers with usual multiplication, [R − 1] be the set of all real numbers minus 1, with binary operation * defined by
$$a * b = a + b - ab.$$

Example 25: Let $G = \{1. -1, i, -i\}$ be a set with usual multiplication of complex numbers. G is a subset of complex numbers, therefore multiplication is associative and commutative. Number 1 is identity element of the set.

Closure and inverse can be seen from following table:

	1	−1	i	−i
1	1	−1	i	−i
−1	−1	1	−i	i
i	i	−i	−1	1
−i	−i	i	1	−1

[Number at intersection of column of top row number and row of left most column number is multiplication of the two numbers]

| Number | 1 | −1 | i | −i |
| Inverse | 1 | −1 | −i | i |

Therefore $G = \{1. -1, i, -i\}$ is an abelian group.

Example 26: Set of all non-singular 2×2 matrices under usual multiplication is an abelian group, since

(i) Matrix multiplication is associative

(ii) $\begin{pmatrix} 1 & 0 \\ 0 & 1 \end{pmatrix}$ is identity matrix.

(iii) Matrices are non-singular therefore their inverse exist and is 2×2.

[Hence an abelian group]

Consider the set of all permutation of first n natural numbers. For example take first 7 numbers and all permutations of $\{1, 2, 3, 4, 5, 6, 7\}$. One of such permutation is denoted by $f\begin{pmatrix} 1,2,3,4,5,6,7 \\ 3,7,1,5,2,4,6 \end{pmatrix}$.

Example 27: Let two permutation be $f\begin{pmatrix} 1,2,3,4,5,6,7 \\ 3,7,1,5,2,4,6 \end{pmatrix}$ and $g\begin{pmatrix} 1,2,3,4,5,6,7 \\ 3,2,5,7,6,4,1 \end{pmatrix}$.

Further let composition be denoted by * and be defined as $g * f = h\begin{pmatrix} 1,2,3,4,5,6,7 \\ 5,1,3,6,2,7,4 \end{pmatrix}$

This shows that this is also a permutation of $\{1,2,3,4,5,6,7\}$.

Associative condition can be checked easily.

Identity permutation is $I\begin{pmatrix} 1,2,3,4,5,6,7 \\ 1,2,3,4,5,6,7 \end{pmatrix}$

Inverse of $g\begin{pmatrix} 1,2,3,4,5,6,7 \\ 3,7,1,5,2,4,6 \end{pmatrix}$ is $g^{-1}\begin{pmatrix} 1,2,3,4,5,6,7 \\ 3,5,1,6,4,7,2 \end{pmatrix}$

$f * g = \varphi\begin{pmatrix} 1,2,3,4,5,6,7 \\ 1,7,2,6,4,5,3 \end{pmatrix} \neq g * f = h\begin{pmatrix} 1,2,3,4,5,6,7 \\ 5,1,3,6,2,7,4 \end{pmatrix}$

Therefore composition of two permutations is not commutative.

Set of all such permutations is non-abelian group.

Example 28: Let $M_{2\times 2}$ be the set of all 2×2 non-singular matrices Let composition be as usual matrix multiplication, then $M_{2\times 2}$ is a group, but not abelian, because matrix multiplication is not commutative.

(i) Obviously matrix multiplication is associative.

(ii) Unit matrix $\begin{pmatrix} 1 & 0 \\ 0 & 1 \end{pmatrix}$ is an identity matrix

(iii) Every non-singular matrix has inverse, hence has multiplicative inverse Hence $M_{2\times 2}$ is a group, but not abelian

Addition modulo n: Let a and b be any two integers and n is a fixed positive integer. The addition modulo n is the binary composition of a and b written as $(a +_n b)$ and defined as the least non-negative remainder r, when $a + b$ is divided by n.

For example $5 +_7 6 = (5 + 6 = 11 = 7 \times 1 + 4) = 4$

Multiplication modulo n: Let a and b any two integers and n is a fixed positive integer. The multiplication modulo n is the binary composition of a and b written as $a \times_n b$ and defined as the least non-negative remainder r, when ab is divided by n.

For example $5 \times_7 6 = (5 \times 6 = 30 = 7 \times 4 + 2) = 2$

Example 29: Show that the set $G = \{0, 1, 2, 3, 4, 5, 6\}$ is a finite abelian group under addition modulo 7.

The following table shows that addition modulo 7 is closed, associative, commutative, and 0 element is identity element.

$+_7$	0	1	2	3	4	5	6
0	0	1	2	3	4	5	6
1	1	2	3	4	5	6	0
2	2	3	4	5	6	0	1
3	3	4	5	6	0	1	2
4	4	5	6	0	1	2	3
5	5	6	0	1	2	3	4
6	6	0	1	2	3	4	5

Further additive inverse of each number of top line can be looked into left column numbers, where intersection is 0 number in the table.

Pairwise it is as below:

Number	0	1	2	3	4	5	6
Inverse	0	6	5	4	3	2	1

Therefore all the conditions of an abelian group are satisfied.

Example 30: Show that the set $G = \{1, 2, 3, 4, 5, 6\}$ is a finite abelian group under multiplication modulo 7.

The following table shows that multiplication modulo 7 is closed, associative, commutative, and 1 is identity element.

Matrices and Algebraic Structure

$\times_7$	1	2	3	4	5	6
1	1	2	3	4	5	6
2	2	4	6	1	3	5
3	3	6	2	5	1	4
4	4	1	5	2	6	3
5	5	3	1	6	4	2
6	6	5	4	3	2	1

line can be looked into left column numbers, where intersection is 1 in the table. Pairwise it is as below:

Number	1	2	3	4	5	6
Inverse	1	4	5	2	3	6

Multiplicative group of integers modulo m, where m is prime

The set G of $(m-1)$, integers $1, 2, 3, 4, \ldots, m-1$; m being prime is a finite abelian group of order $(m-1)$, composition being multiplication modulo m. minus 1, with binary operation $*$ defined by $a * b = a + b - ab$ is a group.

Some Preliminary Lemmas.

Let G be a group with binary operator $*$ defined on it.

   (i) Identity element is unique

  (ii) Every element $x \in G$ has a unique inverse in G

 (iii) Cancellation law holds.

 (iv) For every $x \in G$ $(x^{-1})^{-1} = x$

  (v) For all $x, y \in G$ $(x * y)^{-1} = y^{-1} * x^{-1}$

Proof: (i) Let e_1, e_2 be two distinct zero element of a group G

$e_1 = e_1 + e_2$ since e_2 is identity element

$e_1 + e_2 = e_2$ since e_1 is identity element.

Therefore $e_1 = e_1 + e_2 = e_2$.

Identity element are not distinct. Hence it is unique.

  (ii) Let y, z be two inverses of some element $x \in G$. Further let e be identity element of G.

$y = e * y = (z * x) * y = z * (x * y) = e * z = z$, since $x * y = e = z * x$.

Hence $y = z$, Inverse is unique.

 (iii) Cancellation law for all $x, y, z \in G$,

let $x * z = y * z$, $x * z * z^{-1} = y * z * z^{-1}$, $x * e = y * e$

$\Rightarrow x = y$

(iv) $x^{-1} * (x^{-1})^{-1} = e = x^{-1}x$, **by cancellation law** $(x^{-1})^{-1} = x$

(v) $(x * y) * (y^{-1} * x^{-1}) = x * (y * y^{-1}) * x^{-1} = x * e * x^{-1} = x * x^{-1} = e$.
therefore $(x * y)^{-1} = y^{-1} * x^{-1}$

Example 31: Set $G = \{1, -1\}$ with usual multiplication as binary operation..

(i) There are only two elements, associativity is not required.

(ii) 1 is identity element

1 is inverse of 1 and -1 is inverse of -1.

$G = \{1, -1\}$ is a group.

Order of an element: Let $a \in G$ and e be identity element of the group. If $a^n = e$, where n is an non negative integer then n is called order of a. If n is finite then order is finite otherwise infinite.

Subgroup: Let $(G, *)$ be a group and S be a subset of $(G, *)$. If S with same operation $*$ is a group within itself, then $(S, *)$ is known as subgroup of G.

Example 32: Set of all integers is a subgroup of set of all real numbers with usual addition.

Example 33: Set of four matrices $\begin{pmatrix} 1 & 0 \\ 0 & 1 \end{pmatrix}, \begin{pmatrix} -1 & 0 \\ 0 & 1 \end{pmatrix}, \begin{pmatrix} 1 & 0 \\ 0 & -1 \end{pmatrix}, \begin{pmatrix} -1 & 0 \\ 0 & -1 \end{pmatrix}$ forms a subgroup of $M_{2 \times 2}$ the set of all 2×2 non-singular matrices with composition as usual matrix multiplication

Lemma: A nonempty subset H of a group G with binary operation $*$ is a subgroup if and only if:

1. $x, y \in H \Rightarrow x * y \in H$
2. $x \in H \Rightarrow x^{-1} \in H$.

Proof: If H is subgroup of a group G with binary operation $*$, then conditions 1 and 2 follow. Now, if 1 and 2 are satisfied, then to prove that H is a subgroup of G.

(i) Associability is inherited property

(ii) Condition 1 satisfies closure property.

(iii) Condition 2 gives $x * x^{-1} \in H \Rightarrow e \in H$, since $x * x^{-1} = e$ identity element.

All conditions of group are satisfied Hence proved.

Cyclic Groups: A group G is called a cyclic group, if it is generated by a single element a of G i.e., every element of G is of the form $x = a^m$, where m is some integer. Element a is called generator of the group.

Example 34: Set $G = \{1, -1, i, -1\}$ with usual multiplication of complex numbers is a group.

$i = (i)^1, -1 = (i)^2, -i = i^3, 1 = (i)^4$. therefore group G is a cyclic group and order of the group is 4. Generator is i.

Some properties of Cyclic Groups.

(a) Every cyclic group is abelian
(b) If a is an element of group G such that order of a is n, then order of the cyclic group generated by a is also n.
(c) Order of a finite cyclic group is same as the order of its generator.
(d) If a is a generator of cyclic group. Then a^{-1} is also a generator of the group.
(e) Every subgroup of a cyclic group is cyclic.
(f) Every infinite cyclic group has two and only two generators.

Proof: (a) Let G be a cyclic group. Let $x = a^m$, $y = a^p$, where a is generator of order n. $x, y \in G$
$x * y = a^m * a^p = a^p * a^m = y * x$, hence commutative.

(b) Order of a is n i.e., $a^n = e$ identity element. Let $x \in G$ be generated by a, then elements of G are a^m for integer value $m = 1, 2, 3,....(n-1), n$. This shows that number of elements is n, so order of G is n.

(c) Order of a finite cyclic group is same as the order of its generator follows from (a) and (b).

(d) Let a be a generator of group G i.e., $(G,*) = [a]$. Since a is generator of G, then a^{-1} is also generator of G. because of negative integer power

(e) Let S be a subgroup of group G. Let $x, y \in S$ then for some integer n_1, n_2 $x = a^{n_1}$, $y = a^{n_2}$, where a is a generator of G. $x * y = a^{n_1 + n_2} \in S$ because S is a subgroup

(f) Every infinite cyclic group has two and only two generators. From part (d) above, if a is a generator, then a^{-1} is also a generator. Suppose a^m is also another generator $m \neq \mp 1$, then $a = (a^m)^n = a^{mn}$.

Now $aa^{-1} = a^{mn}a^{-1}$, since $aa^{-1} = e =$ identity element, this shows that order of a is $(mn - 1)$ which is finite, this contradicts that group is infinite. Hence a^m can not be a generator except $m = \mp 1$. Therefore there are only two generators.

Cosets: Let G be a multiplicative group and H be subgroup of G. Let $a \in G$, then the set $Ha = \{ha: h \in H\}$ is said to be right coset of H in G generated by a.

Similarly $aH = \{ah: h \in H\}$ is said to be left coset of H in G generated by a.

If group G is abelian, then cosets $aH = Ha$ for all $a \in G$.

Example 35: Let Q be the set of all integers. Further let $\{....-8, -4, 0, 4, 8, 12.....\}$ be a subgroup under binary operation usual addition.

For $0 \in G$, coset of $H = \{-8, -4, 0, 4, 8, 12,...\}$,
For $1 \in G$, coset of $H = \{-11, -7, -3, 1, 5, 9,...\}$,
For $2 \in G$, coset of $H = \{-10, -6, -2, 2, 6, 10,...\}$,
For $3 \in G$, coset of $H = \{-9, -5, -1, 3, 7, 11,...\}$,
These are only 4 disjoint cosets.

Properties of cosets: Any two right (left) cosets of a subgroup G are either identical or disjoint.

Lagrange's Theorem: Order of each subgroup of a finite group is a divisor of the order of the group.

Proof: Let H be a subgroup of a finite group G such that $(H) = m$, and $O(G) = n$. Let $h_1, h_2, h_3, \ldots, h_m$, be m distinct elements of H.
Let $a \in G$, then Ha is a right coset of H in G. We have $Ha = \{h_1 a, h_2 a, h_3 a, \ldots, h_m a\}$. Ha has m distinct elements

It is known that any two right cosets are either identical or disjoint. Let distinct right cosets of H in G be p.

$G = Ha_1 \cup Ha_2 \cup \ldots \cup Ha_p$, assuming that $(m - p)$ are identical to some of the remaining. $Q(G) = Q(Ha_1) + Q(Ha_2) + \cdots + Q(Ha_p)$

$O(G) = m + m + \cdots + m$; p times
$= pm$.

Therefore $n = pm, \Rightarrow p = \dfrac{n}{m}$. Hence m is divisor of n.

Remark: Converse Lagrange's theorem is not necessarily true.

Every group of prime order is cyclic.

Normal Subgroups: Let $(G,*)$ be a group and $(H,*)$ be a subgroup of $(G,*)$. $(H,*)$ is called Normal subgroup of $(G,*)$ if and only if for every $x \in G$ and all $h \in H$, $xhx^{-1} \in H$.

Simple Group: A group having improper normal subgroup is said to be simple group. For example a group of prime order is a simple group.

Properties of normal subgroup.

(a) Every subgroup of an abelian group is normal subgroup.
(b) Every subgroup of a cyclic group is normal subgroup.
(c) A subgroup H of a group G is a normal subgroup if and only if for every x in G $xHx^{-1} = H$.
(d) Intersection of any two normal subgroup is a normal subgroup.
(e) A subgroup H of a group G is anormal subgroup of G iff the product of two left (right) cosets of H in G ia again a left (right) cosets of H in G.
(f) A subgroup H of a group G is a normal subgroup of G if and only if every left coset of H in G is a right coset of H in G.
(g) If H is a subgroup of a group G such that $x^2 \in H \; \forall \; x \in G$, then H is normal subgroup of G.

Proof: (a) Let H be a subgroup of an abelian group G. For every $x \in G$ and all $h \in G$, $he \in H$, $hxx^{-1} \in H$, $xhx^{-1} \in H$ since operation is commutative. Therefore $xhx^{-1} \in H$. Every subgroup of an abelian group is normal subgroup.

(b) Let H be a subgroup of a cyclic group G Let $h \in H$, $xhx^{-1} = a^m a^n a^{-m}$ for some m, n, where a is a generator of G. $xhx^{-1} = a^m a^n a^{-m} = a^{m+n-m} = a^n = h \in H$. Therefore every subgroup of a cyclic group is normal.

(c) If $xHx^{-1} = H$ then $xhx^{-1} \in H$ for all $x \in G$, H is a normal subgroup. Now suppose H is a normal subgroup of G, then for all $h \in H$ and $x \in G$ by definition $xhx^{-1} \in H$ for all $h \in H$, $xHx^{-1} \subseteq H$, now if for $h \in H$, $xhx^{-1} \notin H$, therefore H is not a normal subgroup, which is a contradiction. $xhx^{-1} \notin H$ not true, so $xhx^{-1} \notin H$., $H \subseteq xHx^{-1}$.

Hence $xHx^{-1} = H$.

(d) Let H_1 and H_2 be two normal subgroup of a group G. Further let $h \in H = H_1 \cap H_2$. $xhx^{-1} \in H_1, H_2$. Since H_1 and H_2 are normal subspaces. $xhx^{-1} \in H_1 \cap H_2$. Hence intersection of any two normal subgroup is a normal subgroup.

(e) Let aH be a left coset in G for some $a \in G$. Consider baH for some $b \in G$. Now $ba = c \in G$. Therefore cH is a left coset of H in G. This proves that multiplication of two left cosets is a again left coset.

(f) If every left coset of H in G is a right coset of H in G, then $ha = ah$ for all $h \in H$ and $a \in G$. $aha^{-1} = haa^{-1} = h \in H$. H is a normal subgroup. Conversely if H is a normal subgroup of G, then $aha^{-1} \in H$. $aha^{-1} = h \Rightarrow ah = ha$

Therefore left coset is a right coset.

(g) If H is a subgroup of a group G such that $x^2 \in H \ \forall \ x \in G$, then H is normal subgroup of G. Let $h = x^2 \in H$ $xhx^{-1} = x \ x^2 \ x^{-1} = x^2 \in H$. Therefore H is a normal subgroup.

Homomorphism and Isomorphism of Groups.

Let G and G_1 be two groups with binary operations $*$ and $\circ$. A mapping $f: G \to G_1$ is called homomorphism if $f(x * y) = f(x) \circ f(y)$ for all $x, y \in G$.

Endomorphism. A homomorphism from G into itself is called endomorphism.

Monomorphism. If a homomorphism from G to G_1 is one to one i.e., for all $x, y \in G$, $f(x) = f(y) \Rightarrow x = y$. then it is called monomorphism,

Epimorphism. If a homomorphism from G to G_1 is onto i.e., for all $z \in G_1$ There exits some $x \in G$ such that $z = f(x)$ then it is called epimorphism.

Isomorphism: If a homomorphism from G to G_1 is one one and onto then it is called isomorphism.

Automorphism. A homomorphism from G onto itself is called automorphism.

Isomorphic groups: Two groups G and G_1 are called isomorphic to each other, if there exists an isomorphism f from G to G_1.

Kernel of a map: Let G and G_1 be two groups with binary operations $*$ and $\circ$. Let mapping $f: G \to G_1$ be homomorphism. Set $X = \{x/f(x) = e_1\}$ is called kernel of f, where e_1 is identity element of G_1.

Example 36: Let set of real numbers $(R, +)$ be a group. Define a map $f: R \to R$ by $f(x) = 2x$.

Let $x_1, x_2 \in R$. $f(x_1 + x_2) \Rightarrow 2(x_1 + x_2) \Rightarrow 2x_1 + 2x_2 = f(x_1) + f(x_2)$.
This shows that f is homomorphism..
$f(x_1) = f(x_2) \Rightarrow 2(x_1) = 2(x_2) \Rightarrow x_1 = x_2$.. This shows that f is one-one.

Let $y \in R$, $y = f\left(\dfrac{y}{2}\right)$, $\dfrac{y}{2} \in R$, therefore f is onto.

Hence mapping is isomorphism because it is one-one and onto.

Example 37: Let set of real numbers $(R, +)$ be a group. Define a map $f: R \to R$ by $f(x) = 0$. Then R set of real numbers is kernel of f.

Example 38: Let R be a group of all integers with usual addition and group $G_1 = (R, *)$ be defined by $a * b = a + b - 1$ for all $a, b \in R$. Define a map $f: R \to G_1$ by $f(x) = x+1$. Prove that t is isomorphism.

Now to check $f(x_1 + x_2) = f(x_1) * f(x_2)$
L.H.S. $= f(x_1 + x_2) = x_1 + x_2 + 1$ by definition
Now $f(x_1) = x_1 + 1$, $f(x_2) = x_2 + 1$
R.H.S.$= f(x_1) * f(x_2) = (x_1 + 1) * (x_2 + 1) = x_1 + 1 + x_2 + 1 - 1 = x_1 + x_2 + 1 = $ L.H.S.
Hence $f(x) = x + 1$ is homomorphism.
$f(x_1) = f(x_2) \Rightarrow x_1 + 1 = x_2 + 1 \Rightarrow x_1 = x_2$ so it is one-one
For any $y \in R$, $y = f(y - 1)$, $y - 1 \in R$, so this is also onto.
Therefore f is isomorphism.

1.6 RINGS

Let S be a nonempty set. Let two operations $(S, +)$ and $(S, \cdot)$ be defined on set S, known as addition and multiplication.

If S satisfy the following conditions, then S is called a ring.
(i) $(S, +)$ is an abelian group
(ii) $(S, \cdot)$ is a semi-group
(iii) $a \cdot (b + c) = a \cdot b + a \cdot c$ and $(a + b) \cdot c = a \cdot c + b \cdot c$.

If $(S, \cdot)$ has multiplicative identity, then it is known as ring with unity.

Matrices and Algebraic Structure

If multiplication is commutative, then it is known as commutative ring.

Restated as: Ring. Let S be a nonempty set. Let two operations $(S, +)$ and $(S, \cdot)$ known as addition and multiplication be defined on set S. If the following conditions are satisfied, then S is called a ring.

(a) Operation + is associative $\quad a + (b + c) = (a + b) + c$ for all $a, b, c \in S$.

(b) Additive identity. If there is an element $0 \in S$ such that $a + 0 = 0 + a = a$ for all $a \in S$.

(c) Existence of additive inverse. If for each $a \in S$, there is an element x in S such that $a + x = x + a = 0$, then x is called inverse of a. x is different for different a.

(d) Operation + is commutative $a + b = b + a$ for all $a, b \in S$

(e) Multiplication is associative $a \cdot (b \cdot c) = (a.b).c$

(f) Multiplication is distributive over addition i.e.,
$$a \cdot (b + c) = a \cdot b + a \cdot c \text{ and } (a + b) \cdot c = a \cdot c + b \cdot c$$

Ring with unity: If a ring has unity i.e., (g) There exist an multiplicative unit element denoted by 1, $1 \in S$, such that $1 \cdot a = a \cdot 1 = a$ for all $a \in S$.

Commutative Ring (h) If S is commutative with respect to operation; then Ring is known as commutative ring.

Example 39: Set of all integers with usual addition and multiplication is commutative ring with unity.

Example 40: Let Z be set of all integers. Two operations $\oplus$, $\odot$ be defined by $x \oplus y = x + y - 5$ and $x \odot y = x + y - 2xy$. Check whether $(Z \oplus, \odot)$ is a ring.

Obviously both operations are closed.

(i) $(x \oplus y) \oplus z = (x + y - 5) \oplus z = x + y + z - 10$
$\quad x \oplus (y \oplus z) = x \oplus (y + z - 5) = x + y + z - 10$

This shows that $(x \oplus y) = x \oplus (y \oplus z)$. $\oplus$ is associative

(ii) $x \oplus e = x + e - 5 = x \Rightarrow e = 5$ identity element

(iii) $(x \oplus y) = (x + y - 5 = e = 5) \Rightarrow y = 10 - x$ additive inverse of x

(iv) $x \oplus y = y \oplus x$

For multiplication $\odot$, $x \odot y = x + y - 2xy$, distributive law does not hold.
$x \odot (y \oplus z) = x \odot (y + z - 5) = x + y + z - 5 - 2x(y + z - 5)$
$\quad = x + y + z - 2xy - 2xz + 10x - 5.$
$(x \oplus y) \odot z = (x + y - 5) \odot z = x + y + z - 5 - 2z(x + y - 5)$
$\quad = x + y + z - 2zx - 2zy + 10z - 5.$
$x \odot (y \oplus z) \neq x \oplus y) \odot z$ Distributive law does not hold.

This is not a ring.

Properties of Rings: In a ring $(R,+,\cdot)$,
  (a) The zero element z is unique,
  (b) The additive inverse of each ring element is unique
  (c) Cancellation law of addition. For all $x, y, z \in R$
   $$x + z = y + z \Rightarrow x = y$$
   $$z + x = z + y \Rightarrow x = y$$

Proof: (a) The zero element z is unique.

Let e_1, e_2 be two distinct zero element of the ring R. $e_1 = e_1 + e_2$ since e_2 is identity element.

$e_1 + e_2 = e_2$ since e_1 is identity element.

Therefore $e_1 = e_1 + e_2 = e_2$. Identity element are distinct. Hence it is unique.

(b) The additive inverse of each ring element is unique. for all $x \in R$

Let y, z be two additive inverse of some $x \in R$. Further let e be identity element of R.

$y = e + y = x + z + y = x + y + z = e + z = z$ since $x + y = e = x + z$

Hence $y = z$ Additive inverse is unique.

(c) Cancellation law of addition. For all $z + x = y + z \Rightarrow x = y$

$x + z = z + y \Rightarrow x = y$

Since $-z \in R$ adding z to both sides

$x + z - z = y + z - z \Rightarrow x = y$

Similarly $z + x = z + y \Rightarrow x = y$

Example 41: The set $R = (a, b)$ with multiplication and addition as below;

+	a	b
a	a	b
b	b	a

+	a	b
a	a	b
b	a	a

is a commutative ring with unity, a is its zero element and b is its unit element.

Example 42: Show that the set $F(\{0,1,2,3,4\}, +_5, \times_5)$ is a commutative ring under addition modulo 5 and multiplication modulo 5..

The following table shows that $F(\{0,1,2,3,4\}, +_5, \times_5)$ addition modulo 5 is closed, associative, commutative, 0 elements is identity element.

$+_5$	0	1	2	3	4
0	0	1	2	3	4
1	1	2	3	4	0
2	2	3	4	0	1
3	3	4	0	1	2
4	4	0	1	2	3

Numbers on top row	0	1	2	3	4
Additive Inverse	0	4	3	2	1

For multiplication

$\times_5$	0	1	2	3	4
0	0	0	0	0	0
1	0	1	2	3	4
2	0	2	4	1	3
3	0	3	1	4	2
4	0	4	3	2	1

This shows that above is a commutative ring.

1.7 FIELD

If a commutative ring with unity has multiplicative inverse of each non-zero element, then it is known a field.

In details, If a non-empty set F with two closed operations known as addition and multiplication, defined on it is known as a field, if the following conditions are satisfied.

Let two operations $(F,+)$ and $(F,\cdot)$ known as addition and multiplication be defined on set F and be closed in F.

(a) Operation $+$ is associative $a + (b + c) = (a + b) + c$ for all $a,b,c \; \forall \; F$

(b) Additive identity. If there is an element $a \in F$ such that $a + 0 = 0 + a = a$ for all $a \in F$, 0 is known as additive identity.

(c) Existence of additive inverse. If for each $a \in F$ there is an element x in F such that $a + x = x + a = 0$, then x is called inverse of a. Element x is different for different a.

(d) Operation $+$ is commutative $a + b = b + a$ for all $a, b \in F$.

(e) Multiplication is distributive over addition i.e.,
   $a.(b + c) = a.b + a.c$ and $(a + b).c = a.c + b.c$ for all $a, b, c \in F$.
(f) There exist an multiplicative unit element denoted by $1 \in F$, such that $a.1 = 1. a = a$ for all $a \in F$.
(g) Multiplication is commutative $a \cdot b = b \cdot a$ for all $a, b \in F$.
(h) Multiplicative inverse of each non-zero element exist.

Example 43: Set R of all real numbers with usual addition and multiplication is a field.

Example 44: Set Q of all Rational numbers with usual addition and multiplication is a field.

Example 45: The set of real numbers of the form $(a + b\sqrt{2})$, where a and b are integers with ordinary addition and multiplication forms a ring. Is it a field?

$(a_1 + b_1\sqrt{2}) + (a_2 + b_2\sqrt{2}) = (a_1 + a_2) + (b_1 + b_2)\sqrt{2}$, addition is closed

$(a_1 + b_1\sqrt{2})(a_2 + b_2\sqrt{2}) = (a_1 b_2 + 2a_2 b_1) + (a_1 b_2 + a_2 b_1)\sqrt{2})$, multiplication is closed

(a) Operation + is associative $\{(a_1 + b_1\sqrt{2}) + (a_2 + b_2\sqrt{2})\} + (a_3 + b_3\sqrt{2})$
   $= (a_1 + a_2 + a_3) + (b_1 + b_2 + b_3)\sqrt{2}) = (a_1 + a_2) + \{(b_1 + b_2\sqrt{2}) + (a_3 + b_3\sqrt{2})\}$,

(b) Additive identity. Zero is a additive identity.

(c) Existence of additive inverse. $(-a - b\sqrt{2})$ is additive inverse of $(a + b\sqrt{2})$.

(d) Operation + is commutative. $(a_1 + b_1\sqrt{2}) + (a_2 + b_2\sqrt{2}) = (a_2 + b_2\sqrt{2}) + (a_1 + b_1)\sqrt{2})$

(e) Multiplication is distributive over addition i.e.,
   $(a + b\sqrt{2})[(a_1 + b_1\sqrt{2}) + (a_2 + b_2\sqrt{2})] = (a + b\sqrt{2})(a_1 + b_1\sqrt{2}) + (a + b\sqrt{2})(a_2 + b_2\sqrt{2})$

(f) There exist an multiplicative unit element which is $(1 + 0\sqrt{2})$.

(g) Multiplication is commutative
   $(a_1 + b_1\sqrt{2})(a_2 + b_2\sqrt{2}) = (a_1 b_2 + 2b_1 b_2) + (a_1 b_2 + a_2 b_1)\sqrt{2}) = (a_1 + b_2\sqrt{2}) + (a_1 + b_1\sqrt{2})$,

Therefore above is commutative ring with unity.

(h) Multiplicative inverse of each non-zero element $(a + b\sqrt{2})$ is
$\left(\dfrac{a}{a^2 - 2b^2}\right) + \left(\dfrac{-b}{a^2 - 2b^2}\right)\sqrt{2}$

Which is not necessarily in the given set because $\left(\dfrac{a}{a^2 - 2b^2}\right)$ and $\left(\dfrac{-b}{a^2 - 2b^2}\right)$ may not be integers.

Therefore, above set with given operations is not a field.

EXERCISES SET – 1

1. Given the matrix $A = \begin{bmatrix} 3 & 1 & -5 & 7 \\ 2 & -5 & 3 & 8 \\ -4 & 2 & 6 & 7 \end{bmatrix}$. Write matrix $4A$ and $-3A$.

2. Given $B = \begin{bmatrix} 2 & 4 & -3 \\ 5 & -4 & 7 \\ 6 & 1 & 3 \\ -1 & 3 & 2 \end{bmatrix}$ is 4×3 matrix, write $5B$ and $-2B$.

3. Let I be identity Matrix or unit matrix. Write $6I$ and $-3I$.

4. Add two matrices $A = \begin{bmatrix} 3 & -1 & 3 & 1 \\ 5 & -4 & 0 & 4 \\ 6 & 0 & 0 & -2 \\ -4 & 12 & 5 & -2 \end{bmatrix}$ and $B = \begin{bmatrix} 3 & 5 & 6 & 7 \\ 3 & -4 & 3 & 0 \\ -2 & 0 & 0 & 2 \\ 1 & 4 & 6 & 5 \end{bmatrix}$

5. Let $A = \begin{bmatrix} 2 & 5 & -3 & 4 \\ 3 & -2 & 1 & 2 \\ 1 & 3 & 4 & -2 \end{bmatrix}$ and $B = \begin{bmatrix} 1 & 3 & 2 \\ -3 & 2 & 1 \\ 2 & 1 & -3 \\ 4 & -5 & 1 \end{bmatrix}$ Write AB. Is BA is defined? If defined, is $AB = BA$?

6. Let $A = \begin{bmatrix} 3 & 3 & -4 \\ -4 & 1 & 2 \\ 4 & -5 & 1 \end{bmatrix}$ and $B = \begin{bmatrix} 2 & -3 & 4 \\ 4 & 1 & -2 \\ -4 & 5 & 3 \end{bmatrix}$. Find AB and BA. Is $AB = BA$.

7. Given $A = \begin{bmatrix} 3 & -1 & 3 & 1 \\ 5 & -4 & 0 & 4 \\ 6 & 0 & 0 & -2 \\ -4 & 12 & 5 & -2 \end{bmatrix}$ and $B = \begin{bmatrix} 3 & 5 & 6 & 7 \\ 3 & -4 & 3 & 0 \\ -2 & 0 & 0 & 2 \\ 1 & 4 & 6 & 5 \end{bmatrix}$

Write transpose of these matrices.

8. Let $A = \begin{bmatrix} 2 & -3 & 0 & 5 \\ 4 & 6 & -2 & 1 \\ 5 & 2 & 3 & 4 \end{bmatrix}$, then $A' = \begin{bmatrix} 2 & 4 & 5 \\ -3 & 6 & 2 \\ 0 & -2 & 3 \\ 5 & 1 & 4 \end{bmatrix}$.

Check whether $(A \pm B)' = A' \pm B'$ and also $(A')' = A$.

9. Complete the matrices $A = \begin{bmatrix} 3 & 1 & -2 & 4 \\ & 2 & 3 & 6 \\ & & -4 & -3 \\ & & & 5 \end{bmatrix}$ and $B = \begin{bmatrix} 3 & & & \\ 1 & 2 & & \\ 2 & 3 & 4 & \\ 4 & 6 & -3 & 3 \end{bmatrix}$ such that new matrices thus formed are symmetric and skew-symmetric

10. Given $A = \begin{bmatrix} 2-3i & 3+i & 1-4i \\ 3+2i & 4-3i & 2+i \\ 6 & 5i & 1-i \end{bmatrix}$ and $B = \begin{bmatrix} 2+3i & 3-i & 1+4i \\ 3-2i & 4+3i & 2-i \\ 6 & -5i & 1+i \end{bmatrix}$

Compute AB and BA and check whether $AB = BA$.

11. Show that $A = \begin{bmatrix} 3 & 2-4i & 4+2i & 5-3i \\ 2+4i & -4 & 6-i & 3+2i \\ 4-2i & 6+i & 2 & 1-5i \\ 5+3i & 3-2i & 1+5i & 5 \end{bmatrix}$ is Hermitian.

12. Show that $A = \begin{bmatrix} 2i & 2+4i & -4-2i & 5-3i \\ -2+4i & 0 & 6+i & 3+2i \\ 4-2i & -6+i & -3i & 1+5i \\ -5-3i & -3+2i & -1+5i & 4i \end{bmatrix}$ is Skew-Hermitian.

13. If $A = \begin{bmatrix} 2-3i & -1+3i \\ -5i & 4+2i \end{bmatrix}$, show that $A^\theta A$ is Hermitian matrix

14. Define Unitary matrix. Show that the matrix $\begin{bmatrix} \frac{1}{2}+i\frac{\sqrt{73}}{12} & -\frac{1}{4}+i\frac{1}{3} \\ \frac{1}{4}+i\frac{1}{3} & \frac{1}{2}-i\frac{\sqrt{73}}{12} \end{bmatrix}$

    is a Unitary matrix,

15. Show that the set of all positive rational numbers forms a abelian group under composition * defined as $a * b = \frac{ab}{4}$ for all $a, b \in Q^+$.

16. Show that set of six transformations, $f_1, f_2, f_3, f_4, f_5, f_6$, on the set of complex numbers defined by $f_1(z) = z$, $f_2(z) = \frac{1}{z}$, $f_3(z) = 1 - z$, $4(z) = \frac{z}{z-1}$, $f_5(z) = \frac{1}{1-z}$, $f_6(z) = \frac{z-1}{z}$

    forms a finite non abelian group for composite function.

17. Define the order of the group. Show that the set of all even integers with zero is an abelian group with respect to usual addition.

18. Let $G = \{\omega_1, \omega_2, \omega_3, \ldots \omega_n\}$, be the set of n roots of unity. Show that G is a finite abelian group with unity under usual multiplication.

19. Show whether set $G = \{0, 1, 2, 3, 4, 5, 6, 7\}$ forms a group under addition modulo 8.

20. Show whether set $G = \{1, 2, 3, 4, 5, 6, 7\}$ forms a group under multiplication modulo 8.

21. What is the degree of the permutation [4, 1, 5, 2, 6, 3, 7].

22. Let $f = \begin{pmatrix} 1 & 2 & 3 & 4 & 5 & 6 \\ 6 & 5 & 4 & 3 & 1 & 2 \end{pmatrix}$. Find whether f is even or odd permutation.

23. Let $f = \begin{pmatrix} 1 & 2 & 3 & 4 & 5 \\ 2 & 4 & 5 & 1 & 3 \end{pmatrix}$, $g = \begin{pmatrix} 1 & 2 & 3 & 4 & 5 \\ 1 & 3 & 5 & 4 & 2 \end{pmatrix}$ be the permutation of the set $X = (1\ 2\ 3\ 4\ 5)$.

    Write gf and fg. Are both the same?

24. Show by taking two permutations of $X = (1\ 2\ 3\ 4\ 5)$. that the multiplication of permutations is not the same in general.

25. Write permutation $\begin{pmatrix} 1 & 2 & 3 & 4 & 5 & 6 & 7 \\ 1 & 3 & 2 & 4 & 5 & 7 & 6 \end{pmatrix}$ in product form.

26. Let G be a group. H is a subgroup of G. Let for $x \in G$, $xHx^{-1} = \{xhx^{-1} : h \in H\}$. Prove that xHx^{-1} is a subgroup of G.
27. Prove that the group $G = (\{0, 1, 2, 3, 4\}, +_5)$ is a cyclic group with generators 1 and 4.
28. Prove that if a is an element of a group G such that the order of a is n, then the order of the cyclic subgroup generated by a is also n.
29. Prove that the set of n n^{th} roots of unity is a finite cyclic group of order n.
30. Let Q be the set of all integers. Further let $H = \{\ldots -15, -10, -5, 0, 5, 10, 15\ldots \ldots\}$ be a subgroup under binary operation usual addition. Write all distinct cosets of H in G.
31. Show that the set of inverses of the elements of a right cosets is a left coset i.e., $(Ha)^{-1} = a^{-1}H$
32. Prove that the multiplicative group $\{1, -1, i, -i\}$ and $(\{0, 1, 2, 3\}, +_4)$ are isomorphic to each other.
33. Show that any two cyclic group of the same order are isomorphism.
34. If H is a subgroup of G and N is a normal subgroup of G, then prove that $H \cap N$ is a normal subgroup of H.
35. Prove that set of all ordered pairs (a, b) of real numbers is a field under addition ÷ and multiplication · defined as
    $(a,b) + (c,d) = (a + c, b + d)$
    $(a,b) \cdot (c,d) = (ac - bd, bc + ad)$
36. Show that the set G of all real valued continuous functions defined on the closed interval $[0,1]$ is a commutative ring with respect to point wise addition and point wise multiplication of functions defined as
    $(f + g)(x) = f(x) + g(x)$
    $(fg)(x) = f(x)$ for all $f, g \in G$

ANSWERS TO EXERCISE – 1

1. $4A = \begin{bmatrix} 12 & 4 & -20 & 28 \\ 8 & -20 & 12 & 32 \\ -16 & 8 & 24 & 28 \end{bmatrix}$, $-3A = \begin{bmatrix} -9 & -3 & 15 & -21 \\ -6 & 15 & -9 & 24 \\ 12 & -6 & -18 & -21 \end{bmatrix}$.

2. $5B = \begin{bmatrix} 10 & 20 & -15 \\ 25 & -20 & 35 \\ 30 & 5 & 15 \\ -5 & 15 & 10 \end{bmatrix}$, $-2B = \begin{bmatrix} -4 & -8 & 6 \\ -10 & 8 & -14 \\ -12 & -2 & -6 \\ 2 & -6 & -4 \end{bmatrix}$.

3. $6I = \begin{bmatrix} 6 & 0 & 0 & 0 \\ 0 & 6 & 0 & 0 \\ 0 & 0 & 6 & 0 \\ 0 & 0 & 0 & 6 \end{bmatrix}$, ... $-3I = \begin{bmatrix} -3 & 0 & 0 & 0 \\ 0 & -3 & 0 & 0 \\ 0 & 0 & -3 & 0 \\ 0 & 0 & 0 & -3 \end{bmatrix}$

4. $A + B = \begin{bmatrix} 6 & 4 & 9 & 8 \\ 8 & -8 & 3 & 4 \\ 4 & 0 & 0 & 0 \\ -3 & 16 & 11 & 3 \end{bmatrix}$.

5. $AB = \begin{bmatrix} -3 & -7 & 22 \\ 19 & -4 & 3 \\ -8 & 23 & -9 \end{bmatrix}$ $BA = \begin{bmatrix} 13 & 5 & 8 & 6 \\ 1 & -16 & 13 & -10 \\ 4 & -1 & -17 & 16 \\ -6 & 33 & -13 & 4 \end{bmatrix}$ $BA \neq AB$

6. $AB = \begin{bmatrix} 34 & -26 & -6 \\ -12 & 23 & -12 \\ -16 & -12 & 29 \end{bmatrix}$ $BA = \begin{bmatrix} 38 & -17 & -10 \\ 0 & 23 & -16 \\ -20 & -22 & 29 \end{bmatrix}$ $BA \neq AB$.

7. $A^T = \begin{bmatrix} 3 & 5 & 6 & -4 \\ -1 & -4 & 0 & 12 \\ 3 & 0 & 0 & 5 \\ 1 & 4 & -2 & -2 \end{bmatrix}$ and $B^T = \begin{bmatrix} 3 & 3 & -2 & 1 \\ 5 & -4 & 0 & 4 \\ 6 & 3 & 0 & 6 \\ 7 & 0 & 2 & 5 \end{bmatrix}$

9. Symmetric Matrices $A = \begin{bmatrix} 3 & 1 & -2 & 4 \\ 1 & 2 & 3 & 6 \\ -2 & 3 & -4 & -3 \\ 4 & -4 & -3 & 5 \end{bmatrix}$ and $B = \begin{bmatrix} 3 & 1 & 2 & 4 \\ 1 & 2 & 3 & 6 \\ 2 & 3 & 4 & -3 \\ 4 & 6 & -3 & 3 \end{bmatrix}$

Skew–Symmetric Matrices $A = \begin{bmatrix} 3 & 1 & -2 & 4 \\ -1 & 2 & 3 & 6 \\ 2 & -3 & -4 & -3 \\ -4 & -6 & 3 & 5 \end{bmatrix}$ and $B = \begin{bmatrix} 3 & -1 & -2 & -4 \\ 1 & 2 & -3 & -6 \\ 2 & 3 & 4 & 3 \\ 4 & 6 & -3 & 3 \end{bmatrix}$

10. Given $AB = \begin{bmatrix} 30-27i & -2-3i & 26+i \\ 18+2i & 41-7i & 7+7i \\ 28+27i & -29i & 13+34i \end{bmatrix}$ and $BA = \begin{bmatrix} 30+27i & -2+3i & 26-i \\ 18-2i & 41+7i & 17-7i \\ 28-27i & -2-9i & 13-34i \end{bmatrix}$

19. Yes **20.** No **21.** 6 **22.** Even **23.** No. **25.** $(123)(67)$

CHAPTER 2

System of Linear Equations

System of linear equations occur frequently in all subjects of Science and Engineering to study various aspects of the physical problems.

Linear equation: An equation of the form $a_1x_1 + a_2x_2 + a_3x_3 + ... + a_nx_n = b$ where $a_1, a_2, ..., a_n$ are some given numbers and $x_1, x_2, ..., x_n$ are unknown variables, is known as linear equation.

For examples

$2x_1 - 3x_2 + 5x_3 = 8$ is a linear equation in x_1, x_2, x_3

and $3x + 4y - 6z + 2w = 18$ is a linear equation in x, y, z and w.

In these equations powers of unknowns is one in each case.

Non-linear equation: If the power of at least one unknown, of the unknowns is other than one, then the equation is known as non-linear.

For examples

$2x_1 - 3x_2^2 + 5x_3 = 10$ is non-linear, since power of x_2 is 2.

$x - y + z^{\frac{3}{2}} = 16$ is non-linear, since power of z is $\frac{3}{2}$.

The aim of this chapter is to learn consistency, existence, uniqueness of solution of linear equations and efficient methods to solve the system of linear equations. A method is called efficient, if it takes comparatively less time. Crammer's rule of finding the solution of a system of linear equations involves evaluation of determinant values of many matrices. It is known that computing the determinant value of a matrix on computer takes more time than solving the system by elimination method; therefore Crammer's rule is not suitable for computation.

2.2 Elementary Linear Algebra

We discuss an integrated approach, which deals with above all aspects of a given system, i.e., namely algorithm of finding the solution, consistency, existence, uniqueness and the solution of the problem involved.

2.1 SOLUTION BY GRAPHS AND ELEMENTARY ROW OPERATIONS

Consider the general system of m linear equations, in n unknowns, $x_1, x_2, ..., x_n$

$$
\begin{aligned}
a_{11}x_1 + a_{12}x_2 + \cdots\cdots + a_{1n}x_n &= b_1 \\
a_{11}x_1 + a_{12}x_2 + \cdots\cdots + a_{1n}x_n &= b_2 \\
\vdots \qquad \vdots \qquad\qquad\qquad \vdots \qquad &\vdots \\
a_{i1}x_1 + a_{i2}x_2 + \cdots\cdots + a_{in}x_n &= b_i \\
\vdots \qquad \vdots \qquad\qquad\qquad \vdots \qquad &\vdots \\
a_{m1}x_1 + a_{m2}x_2 + \cdots\cdots + a_{mn}x_n &= b_n
\end{aligned}
\qquad (1)
$$

This can be written in matrix form as

$$
\begin{bmatrix}
a_{11} & a_{12} & --- & a_{1n} \\
a_{21} & a_{22} & --- & a_{2n} \\
\vdots & \vdots & \vdots & \vdots \\
a_{i1} & a_{i2} & --- & a_{in} \\
\vdots & \vdots & \vdots & \vdots \\
a_{m1} & a_{m2} & --- & a_{mn}
\end{bmatrix}
\begin{bmatrix} x_1 \\ x_2 \\ \vdots \\ \vdots \\ x_n \end{bmatrix}
=
\begin{bmatrix} b_1 \\ b_2 \\ \vdots \\ b_i \\ \vdots \\ b_n \end{bmatrix}
\qquad (2)
$$

On writing A_{mxn} for coefficient matrix, $\mathbf{x} = (x_1, x_2, ..., x_n)^T$ and $\mathbf{b} = (b_1, b_2, ..., b_m)^T$, system in matrix form is $A\mathbf{x} = \mathbf{b}$.

In case of solution of two linear equations in two unknowns, equations can be solved graphically as shown by the following simple cases:

(a) $2x + y = 3$

$4x + 2y = -1$

These two equations are parallel, no point of intersection.

Therefore solution does not exist.

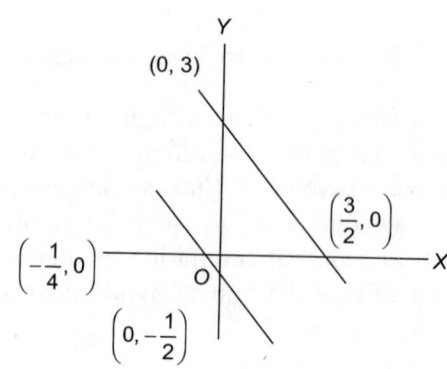

(b) $2x + y = 3$

$4x + 2y = 6$

These are two coinciding lines, having infinite many common points.
Therefore infinite number of solutions exists.

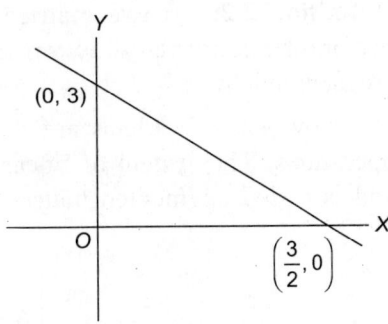

(c) $2x + y = 3$

$x - 2y = -1$

Both lines are passing through the only common point (1, 1).
Hence solution is unique.

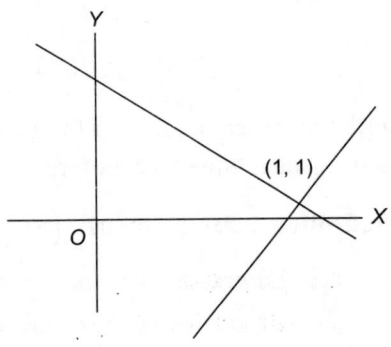

In case of the large system of linear equations, graphical method will not work, therefore non-graphical methods are discussed in details in this chapter.

First some problems of the system of linear equations are being solved by elimination method. These problems will help us to understand the various aspects of the solution of system of linear equations conveniently in general.

Definition 2.1: Elementary row operations: We know that some arithmetic operations done on a system of linear equations do not change the solution. Therefore such following operations are called elementary row operations.

1. Multiplications of any equation (row) by a non-zero number.
2. Interchanging any two equations (rows).
3. Replacing an equation (row) by sum of it self and a multiplication of any other equation (row).

We have to study the procedure of working the problem to see whether the system has
- No solution
- Unique solution
- Infinite many solutions.

In case solution exists, we have to find the solution of the system.

Definition 2.2: If two matrices A and B of same order are obtained from each other by performing elementary row operations, then matrices A and B are called row equivalent and are denoted by $A \sim B$.

Now some problems are solved by elimination method by performing elementary row operations. The system of linear equations (1) can be written in the following matrix form and is called augmented matrix (A, b):

$$\begin{bmatrix} a_{11} & a_{12} & \ldots & a_{1n} & b_1 \\ a_{21} & a_{22} & \ldots & a_{2n} & b_2 \\ \vdots & \vdots & \ldots & \vdots & \vdots \\ a_{m1} & a_{m2} & \ldots & a_{mn} & b_m \end{bmatrix}, \qquad (3)$$

with the understanding of their usual addition of multiplication of coefficients by $x_1, x_2, \ldots, x_n$ and equality sign just before **b**-values.

Definition 2.3: A matrix $(a_{ij})_{m \times n}$ is called in row-echelon form if

(a) All non-zero rows are above all zero rows.

(b) All entries of next rows in a column below non-zero leading entry are zeros.

(c) Each non-zero leading entry of a row in a column is on the right of leading entry of previous row i.e., each non-zero leading entry of row is successively on the right side of the previous row.

Remark: System of linear equations can be solved by bringing the augmented matrix (A, b) to row echelon form only, i.e. reducing to upper triangular form. Later on it would be known that solving the systems from row-echelon form is efficient method i.e. takes less time than if system is solved from row-reduced echelon form (to be discussed in the next section).

System of equations in matrix form $Ax = b$ can also be solved by finding A^{-1}, if $A_{n \times n}$ is invertible i.e. non-singular, then A^{-1} exists and $A^{-1}Ax = A^{-1}b \Rightarrow x = A^{-1}b$, therefore solving the system of linear equations is to find A^{-1} and then $A^{-1}b$, but this is applicable only when A is square invertible matrix A. Finding A^{-1} involves more work than solving by elimination method, therefore A^{-1} method is not preferred.

Definition 2.4: Number of non-zero rows in a row echelon form of a matrix A is called row-rank of A. (equivalently number of non-zero rows in a row-reduced echelon form of a matrix A is called row-rank of A. An equivalent of row-rank of A is given in chapter 3 also).

To analyze the nature of the solution of a system of linear equations, first the following examples are solved by elementary row operations.

System of Linear Equations

Example 1:
$$4x - 2y + 3z = -1$$
$$x + 5y - 2z = 21$$
$$-2x + 3y + 6z = -7$$

Augmented matrix **(A, b)** is denoted by W and called working matrix

On interchanging first and second rows

i.e. $r_1 \leftrightarrow r_2$, $\sim \begin{bmatrix} 1 & 5 & -2 & 21 \\ 4 & -1 & 3 & -1 \\ -2 & 3 & 6 & -7 \end{bmatrix}$, $\sim$ is used for row equivalent.

On subtracting 4 times of first row from second and adding 2 times of first to 3rd row, ($\leftarrow$ is used for replace by)

i.e. $\quad r_2 \leftarrow r_2 - 4r_1, r_3 \leftarrow r_3 + 2r_1 \sim \begin{bmatrix} 1 & 5 & -2 & 21 \\ 0 & -21 & 11 & -85 \\ 0 & 13 & 2 & 35 \end{bmatrix}$.

Similarly on $\quad r_3 \leftrightarrow r_2 \sim \begin{bmatrix} 1 & 5 & -2 & 21 \\ 0 & 13 & 2 & 35 \\ 0 & -21 & 11 & -85 \end{bmatrix}$

on $\quad r_3 \leftrightarrow r_3 + \dfrac{21}{13} r_2 \sim \begin{bmatrix} 1 & 5 & -2 & 21 \\ 0 & 13 & 2 & 35 \\ 0 & 0 & \dfrac{185}{13} & -\dfrac{370}{13} \end{bmatrix}$

on $\quad r_3 \leftarrow r_3 \Big/ \dfrac{185}{13} \sim \begin{bmatrix} 1 & 5 & -2 & 21 \\ 0 & 13 & 2 & 35 \\ 0 & 0 & 1 & -2 \end{bmatrix}$.

Note: In the above matrix all conditions of echelon form are satisfied, therefore above matrix is in echelon form.

Now, from above matrix form, the equations are
$$x + 5y - 2z = 21$$
$$13y + 2z = 35$$
$$z = -2$$

Solving in backward order, we get

$$z = -2$$
$$13y + 2 \times -2 = 35 \Rightarrow 13y = 39, y = 3$$

and
$$x + 5 \times 3 - 2 \times -2 = 21 \Rightarrow x = 2.$$

Hence unique solution of the system is $x = 2, y = 3, z = -2$.

This procedure of solving in backward order is known as backward substitution. It would be discussed later that above procedure of solving system of linear equations is most efficient. The evaluation of $(n + 1)$ determinant values in crammer's rule takes $(n + 1)$ times the time of the time taken by above method, therefore elimination method is most suitable.

Example 2: Consider the system of linear equations

$$3x - 2y + z = 1$$
$$x + 3y - 2z = 9$$
$$2x - y + 3z = 4$$

$$\text{Augmented matrix } (A, b) = \begin{bmatrix} 3 & -2 & 1 & 1 \\ 1 & 3 & -2 & 9 \\ 2 & -1 & 3 & 4 \end{bmatrix}$$

On $r_1 \leftrightarrow r_2$, interchanging first and second rows

$$\sim \begin{bmatrix} 1 & 3 & -2 & 9 \\ 3 & -2 & 1 & 1 \\ 2 & -1 & 3 & 4 \end{bmatrix}$$

$r_2 \leftarrow r_2 - 3r_1$, subtracting 3 times of entries of first row from corresponding entries of second row, and $r_3 \leftarrow r_3 - 2r_1$, we get

$$\begin{array}{c} r_2 \leftarrow r_2 - 3r_1 \\ r_3 \leftarrow r_3 - 2r_1 \end{array} \sim \begin{bmatrix} 1 & 3 & -2 & 9 \\ 0 & -11 & 7 & -26 \\ 0 & -7 & 7 & -14 \end{bmatrix}$$

$r_2 \leftrightarrow r_3$, interchanging r_2 and r_3, we get

$$\sim \begin{bmatrix} 1 & 3 & -2 & 9 \\ 0 & -7 & 7 & -14 \\ 0 & -11 & 7 & -26 \end{bmatrix}$$

on dividing r_2 by -7, e.i..,

$$r_2 \leftarrow -\frac{1}{7}r_2 \sim \begin{bmatrix} 1 & 3 & -2 & 9 \\ 0 & 1 & -1 & 2 \\ 0 & -11 & 7 & -26 \end{bmatrix}$$

Now adding 11 times of second row to r_3, we get

$$r_3 \leftarrow r_3 + 11r_2 \sim \begin{bmatrix} 1 & 3 & -2 & 9 \\ 0 & 1 & -1 & 2 \\ 0 & 0 & -4 & -4 \end{bmatrix},$$

on dividing r_3 by -4,

$$r_3 \leftarrow -\frac{1}{4}r_3 \sim \begin{bmatrix} 1 & 3 & -2 & 9 \\ 0 & 1 & -1 & 2 \\ 0 & 0 & 1 & 1 \end{bmatrix}$$

Similarly $r_1 \leftarrow r_1 - 3r_2$ and $r_3 \leftarrow r_3 + r_2 \sim \begin{bmatrix} 1 & 0 & 0 & 2 \\ 0 & 1 & 0 & 3 \\ 0 & 0 & 1 & 1 \end{bmatrix}$

$$\Rightarrow \begin{bmatrix} 1 & 0 & 0 \\ 0 & 1 & 0 \\ 0 & 0 & 1 \end{bmatrix} \begin{bmatrix} x \\ y \\ z \end{bmatrix} = \begin{bmatrix} 2 \\ 3 \\ 1 \end{bmatrix} \tag{4}$$

we get unique solution $x = 2$, $y = 3$, $z = 1$.

Example 3: Consider the system of linear equations

$$3x - 2y + z = 1$$
$$x + 3y - 2z = 9$$
$$2x - y + 3z = 4$$
$$x + 2y - 2z = 6$$

Note: Number of equations is more than the number of unknowns in this system of equations

Augmented matrix $(A, b) \sim \begin{bmatrix} 3 & -2 & 1 & 1 \\ 1 & 3 & -2 & 9 \\ 2 & -1 & 3 & 4 \\ 1 & 2 & -2 & 6 \end{bmatrix}$

On performing elementary row operations, we get

$$= \begin{bmatrix} 1 & 0 & 0 & 2 \\ 0 & 1 & 0 & 3 \\ 0 & 0 & 1 & 1 \\ 0 & 0 & 0 & 0 \end{bmatrix} \quad (5)$$

Note: All entries in 4th row are zeros.

Now equivalent equations give unique solution $x = 2, y = 3, z = 1$.

Example 4: Consider the system of linear equations

$$3x - 2y + z = 1$$
$$x + 3y - 2z = 9$$
$$2x - y + 3z = 4$$
$$x - 3y + 2z = 3$$

Corresponding augmented matrix (A, b)

$$= \begin{bmatrix} 3 & -2 & 1 & 1 \\ 1 & 3 & -2 & 9 \\ 2 & -1 & 3 & 4 \\ 1 & -3 & 2 & 3 \end{bmatrix}$$

Note: Number of equations is more than number of unknowns.

Applying elementary row operations, we get

$$\sim \begin{bmatrix} 1 & 0 & 0 & 2 \\ 0 & 1 & 0 & 3 \\ 0 & 0 & 1 & 1 \\ 0 & 0 & 0 & 1 \end{bmatrix} \quad (6)$$

Last row gives zero $0z = 1$, which can not be true, therefore solution does not exist. Such a system is called inconsistent.

Example 5: Consider the equations

$$x_1 + 2x_2 - 2x_3 = 8$$
$$x_1 + x_2 - 3x_3 = 9$$

Number of equations is less than number of unknowns.

Augmented Matrix is $\begin{bmatrix} 1 & 2 & -2 & 8 \\ 1 & 1 & -3 & 9 \end{bmatrix}$

on $r_1 \leftrightarrow r_2$ ~ $\begin{bmatrix} 1 & 1 & -3 & 9 \\ 1 & 2 & -2 & 8 \end{bmatrix}$

on $r_2 \leftarrow r_2 - r_1$ ~ $\begin{bmatrix} 1 & 1 & -3 & 9 \\ 0 & 1 & 1 & -1 \end{bmatrix}$

Now $r_1 \leftarrow r_1 - r_2$ ~ $\begin{bmatrix} 1 & 0 & -4 & 10 \\ 0 & 1 & 1 & -1 \end{bmatrix}$ (7)

on writing in the form of equations

$$\left. \begin{array}{c} x_1 - 4x_3 = 10 \\ x_2 + x_3 = -1 \end{array} \right] \Rightarrow \begin{array}{c} x_1 = 4x_3 + 10 \\ x_2 = -x_3 - 1 \end{array}$$

In these equations x_1 and x_2 are expressed in terms of x_3.

So on assigning different values to x_3, we get different sets of values of x_1 and x_2.

Therefore this system has infinite number of solutions.

Remark: From the above examples, we note that the final form of augmented matrices (4), (5), (6) and (7) has some differences in them.

All three rows are non-zero in (4), last row i.e. 4th row is of zeros in (5), in (6) last row i.e. 4th corresponding to coefficient matrix is of zeros, but last row of augmented matrix is not zero i.e. 4 non-zero rows.

These differences have resulted in the different nature of answers in the above problems. Final form of augmented matrices in the above four problems, has some common features, though there are differences among them. These common features are summarized in the next section.

2.2 ROW REDUCED ECHELON FORM

Definition 2.5: If a row echelon matrix A_{mxn} satisfies the following additional conditions, then matrix A is called in row-reduced echelon form.

(a) Non-zero leading entry in a row is 1.

(b) If a column contains leading entry 1, then all other entries in that column are zeros.

(c) If a row contains non-zero entries, next to non-zero leading entry, then all other entries below it in that column are zeroes.

2.10   Elementary Linear Algebra

Example 6: Consider the matrix

$$\begin{bmatrix} 1 & 0 & -1 & 0 & 0 & 2 \\ 0 & 1 & 1 & 0 & 0 & 3 \\ 0 & 0 & 0 & 1 & 0 & 1 \\ 0 & 0 & 0 & 0 & 1 & -2 \\ 0 & 0 & 0 & 0 & 0 & 0 \end{bmatrix}$$

In the first row and first column entry is 1, below all other entries of first column are zeroes. In second row first non-zero entry 1 occurs in second column, and all other entries of second column are zeroes. In second row third column 1 is not leading entry of second row, so non-zero entry -1 is occurring in the first row. In third row non-zero leading entry 1 is appearing in fourth column and all other entries are zeroes in the fourth column. And so on therefore the above matrix is in the row-reduced echelon form, graphically no non-zero entry occurs below steps and in each corner of a step 1 is occuring.

Note: The difference between row-echelon and row reduced echelon form is

(a) Non-zero leading entry in row-reduced echelon form is 1; while in row-echelon, non-zero leading entry may be any non-zero number.

(b) In row echelon form entries of a column, above non-zero leading entry may be any numbers, but in row-reduced echelon form entries of a column above and below non-zero leading entry are zeros.

Further to clarify the difference between row echelon form and row-reduced echelon form, following example is solved by reducing in two different forms.

Example 7: Consider the system of linear equations

$$2x_1 - 3x_2 + 4x_3 = 23$$
$$3x_1 + 4x_2 - 8x_3 = -19$$
$$4x_1 - x_2 - 2x_3 = 11$$

Performing elementary row operations on the augmented matrix

$$(A, b) = \begin{bmatrix} 2 & -3 & 4 & 23 \\ 3 & 4 & -8 & -19 \\ 4 & -1 & -2 & 11 \end{bmatrix},$$

we get

$$\begin{bmatrix} 2 & -3 & 4 & 23 \\ 0 & 8.5 & -14 & -53.5 \\ 0 & 5 & -10 & -35 \end{bmatrix},$$

System of Linear Equations

on $r_2 \leftrightarrow r_3$ ~ $\begin{bmatrix} 2 & -3 & 4 & 23 \\ 0 & 1 & -2 & -7 \\ 0 & 8.5 & -14 & -53.5 \end{bmatrix}$

on $r_3 - 8.5\, r_2$ ~ $\begin{bmatrix} 2 & -3 & 4 & 23 \\ 0 & 1 & -2 & -7 \\ 0 & 0 & 1 & 2 \end{bmatrix}$,

the above matrix is in row echelon form.

Row rank of $A = 3 =$ row rank of (A, b). Hence unique solution exists.

Corresponding equations are $\quad 2x_1 - 3x_2 + 4x_3 = 23$
$$x_2 - 2x_3 = -7$$
$$x_3 = 2$$

On solving in backward order, we get $x_3 = 2$, $x_2 = -3$, $x_1 = 3$

(Known as backward substitution) i.e. $x_1 = 3$, $x_2 = -3$, $x_3 = 2$ unique solutions exists

Example 8: From working matrix of example 7

$$\begin{bmatrix} 2 & -3 & 4 & 23 \\ 0 & 1 & -2 & -7 \\ 0 & 0 & 1 & 2 \end{bmatrix},$$

if we bring to diagonal form

By $r_1 \leftarrow 3r_2 + r_1$, $\quad r_2 \leftarrow 2r_3 + r_2$ ~ $\begin{bmatrix} 2 & 0 & -2 & 6 \\ 0 & 1 & 0 & -3 \\ 0 & 0 & 1 & 2 \end{bmatrix}$

$r_1 \leftarrow 2r_3 + r_1$ ~ $\begin{bmatrix} 2 & 0 & 0 & 6 \\ 0 & 1 & 0 & -3 \\ 0 & 0 & 1 & 2 \end{bmatrix}$

$r_1 \leftarrow r_1/2$ ~ $\begin{bmatrix} 1 & 0 & 0 & 3 \\ 0 & 1 & 0 & -3 \\ 0 & 0 & 1 & 2 \end{bmatrix}$,

this matrix is in row-reduced echelon form.

Note: In first column non-zero leading entry is in first row and other entries in first column are 0. In 2nd row non-zero leading entry 1 in second column zero all other entries are zeros in the second column. In third row non-zero leading entry is 1 in third column and other entries are zeros in third column.

On writing back in equation form, we get the solution the same solution

$$x_3 = 2, \; x_2 = -3, \; x_1 = 3.$$

2.3 CONSISTENCY OF THE LINEAR SYSTEM

Definition 2.6: If no solution of a linear system of equations exists, then the system is called inconsistent. Now we analysis the results of some examples of section 2.1.

In example 2, number of equations $m = 3$, number of unknowns $n = 3$, Row rank of A = Row rank of $(A, b) = 3 = n$, system has a unique solution.

In example 3, number of rows $m = 4$, number of unknowns $n = 3$, Row-rank (A) = row-rank of $(A, b) = 3$, system has unique solution.

In example 4, number of rows $m = 4$, number of unknowns $n = 3$, row-rank of $A = 3$, but row-rank of $(A, b) = 4 \neq$ row-rank of A, system has no solution i.e. system is inconsistent.

In example 5, number of rows $m = 2$, number of unknowns $n = 3$, row-rank of A = row-rank $(A, b) = 2 <$ number of unknowns $n = 3$, the system has infinite many numbers of solutions.

In general

(a) If row-rank of A = row-rank of (A, b), system is consistent and solution exists.

(b) If solution exists i.e. row-rank of A = row-rank of $(A, b) = n$ number of unknowns, then system is consistent and solution is unique.

(c) If row-rank (A) = row-rank $(A, b) < n$ number of unknowns, then system is consistent and system has infinite solutions.

(d) If row-rank of $A \neq$ row-rank of (A, b), then system is inconsistent, solution does not exist.

Now some more examples are being solved for clarity of consistency of the system of linear equations.

Example 9: Consider the system of equations.

$$2x_1 - 3x_2 + 4x_3 = 23$$
$$3x_1 + 4x_2 - 8x_3 = -19$$
$$9x_1 - 5x_2 + 4x_3 = 30$$

$$\text{Augmented matrix } (A, b) = \begin{bmatrix} 2 & -3 & 4 & 23 \\ 0 & 8.5 & -14 & -53.5 \\ 0 & 0 & 0 & -40 \end{bmatrix}$$

Row rank of A = 2 ≠ 3 = Row rank of (A, b). Hence system is inconsistent.

Also writing back in equations, we get $2x_1 - 3x_2 + 4x_3 = 23$, $8.5x_2 - 14x_3 = -30.5$ and $0x_3 = -40$.

Last equation is not valid, therefore above system of linear equation is inconsistent.

Example 10: Check whether the given system of linear equations is consistent.

$$2x_1 - 3x_2 + 4x_3 = 23$$
$$3x_1 + 4x_2 - 8x_3 = -19$$
$$4x_1 - x_2 - 2x_3 = 11$$
$$5x_1 + 6x_2 - 14x_3 = -26$$

Augmented matrix (A, b)

$$= \begin{bmatrix} 2 & -3 & 4 & 23 \\ 3 & 4 & -8 & -19 \\ 4 & -1 & -2 & 11 \\ 5 & 6 & -14 & -26 \end{bmatrix}$$

on applying elementary row operations, we get

$$\begin{bmatrix} 2 & -3 & 4 & 23 \\ 0 & 1 & -2 & 7 \\ 0 & 0 & 1 & 2 \\ 0 & 0 & 0 & -5 \end{bmatrix}$$

We note that row rank of $A = 3 \neq 4 =$ row rank of augmented matrix (A, b).

Therefore the above system is not consistent.

Example 11: Is the following systems of linear equations consistent? If consistent, then find the solutions of the system, use elementary row operations.

$$2x_1 + 3x_2 + 2x_3 = 16$$
$$3x_1 + x_2 + x_3 = 6$$
$$x_1 + 5x_2 + 3x_3 = 1$$

Augmented matrix $(A, b) = \begin{bmatrix} 2 & 3 & 2 & 16 \\ 3 & 1 & 1 & 6 \\ 1 & 5 & 3 & 10 \end{bmatrix}$, on $r_1 \leftrightarrow r_3 \sim \begin{bmatrix} 1 & 5 & 3 & 10 \\ 3 & 1 & 1 & 6 \\ 2 & 3 & 2 & 16 \end{bmatrix}$

on $r_2 \leftarrow r_2 - 3r_1$ and $r_3 \leftarrow r_3 - 2r_1 \sim \begin{bmatrix} 1 & 5 & 3 & 10 \\ 0 & -14 & -8 & -24 \\ 0 & -7 & -4 & -4 \end{bmatrix}$

on $r_2 \leftarrow r_2 - 2r_3, \sim \begin{bmatrix} 1 & 5 & 3 & 10 \\ 0 & 0 & 0 & -16 \\ 0 & -7 & -4 & -4 \end{bmatrix}$, $r_2 \leftrightarrow r_3 \sim \begin{bmatrix} 1 & 5 & 3 & 10 \\ 0 & -7 & -4 & -4 \\ 0 & 0 & 0 & -16 \end{bmatrix}$

$r_2 \leftarrow r_2/-7$ and $r_3 \leftarrow r_3/-16 \sim \begin{bmatrix} 1 & 5 & 3 & 10 \\ 0 & 1 & 0.57143 & 0.57143 \\ 0 & 0 & 0 & 1 \end{bmatrix}$

on $r_1 \leftarrow r_1 - 5r_2, \sim \begin{bmatrix} 1 & 0 & 0.14285 & 7.14 \\ 0 & 1 & 0.57143 & 0.57143 \\ 0 & 0 & 0 & 1 \end{bmatrix}$

row-rank of $A = 2 \neq$ row-rank of $(Ab) = 3$, not equal therefore system is inconsistent.

Example 12: Consider the system linear equations
$$2x_1 - 3x_2 + 4x_3 = 23$$
$$3x_1 + 4x_2 - 8x_3 = -19$$
$$4x_1 - x_2 - 2x_3 = 11$$
$$x_1 + 2x_2 - 2x_3 = -7$$

On applying elementary row operations on augmented matrix (A, b),

we get, working matrix $\begin{bmatrix} 2 & -3 & 4 & 23 \\ 0 & 1 & -2 & -7 \\ 0 & 0 & 1 & 2 \\ 0 & 0 & 0 & 0 \end{bmatrix}$.

Row rank of $(A, b) = 3 =$ row rank of A. Hence system is consistent.
Unique solution $x_1 = 2$, $x_2 = -3$, $x_3 = 3$ exists.

System of Linear Equations

Example 13: Consider
$$2x_1 - 3x_2 + 4x_3 = 23$$
$$3x_1 + 4x_2 - 8x_3 = -19$$
$$5x_1 - x_3 - 4x_3 = 4$$

Augmented matrix $(A, b) = \begin{bmatrix} 2 & -3 & 4 & 23 \\ 3 & 4 & -8 & -19 \\ 5 & -1 & -4 & 4 \end{bmatrix}$

Applying elementary row operation, we get $\begin{bmatrix} 2 & -3 & 4 & 23 \\ 0 & 1 & -2 & -7 \\ 0 & 0 & 0 & 0 \end{bmatrix}$

Row rank of $(A, b) = 2 =$ row-rank of $A \leq 3 =$ the number of unknowns. Hence system has infinite solutions.

Writing back in equations form, we get
$$2x_1 - 3x_2 + 4x_3 = 23$$
$$x_2 - 2x_3 = -7$$

on solving the above equations, we get $x_2 = 2x_3 - 7$

and $\quad\quad 2x_1 - 2x_3 = 2 \Rightarrow x_1 = x_3 + 1$

For various values of x_3, we get different set of values of x_1, x_2.

Therefore system has infinite many solutions.

Example 14: Using elementary row operation, show whether the following system of linear equations is consistent. If consistent, find the solution of the system.
$$x_1 + 2x_2 + x_3 = 2$$
$$2x_1 + x_2 - 10x_3 = 4$$
$$2x_1 + 3x_2 - x_3 = 2$$

Augmented matrix $(A, b) = \begin{bmatrix} 1 & 2 & 1 & 2 \\ 2 & 1 & -10 & 4 \\ 2 & 3 & -1 & 2 \end{bmatrix}$,

on $r_2 \leftarrow r_2 - 2r_1$ and $r_3 \leftarrow r_3 - 2r_1 \sim \begin{bmatrix} 1 & 2 & 1 & 2 \\ 0 & -3 & -12 & 0 \\ 0 & -1 & -3 & -2 \end{bmatrix}$,

$$r_2 \leftarrow r_2/-3 \sim \begin{bmatrix} 1 & 2 & 1 & 2 \\ 0 & 1 & 4 & 0 \\ 0 & -1 & -3 & -2 \end{bmatrix},$$

on $r_1 \leftarrow r_1 - 2r_2,\ r_3 \leftarrow r_3 + r_2 \sim \begin{bmatrix} 1 & 0 & -7 & +2 \\ 0 & 1 & 4 & 0 \\ 0 & 0 & 1 & -2 \end{bmatrix},$

on $r_2 \leftarrow r_2 - 4r_3,\ r_1 \leftarrow r_1 + 7r_3 \sim \begin{bmatrix} 1 & 0 & 0 & -12 \\ 0 & 10 & 0 & 8 \\ 0 & 0 & 1 & -2 \end{bmatrix},$

row-rank of $A = 3 =$ row-rank of $(A, b) =$ number of unknowns and system is consistent and solution is unique. By back substitution $x_3 = -2$, $x_2 = 8$, $x_1 = -12$ i.e. solution vector $x = (-12, 8, -2)^T$.

2.4 NON-SINGULAR MATRICES AND DETERMINANT VALUES

Definition 2.7: A square matrix A_{nxn} is called non-singular if $\det A \neq 0$, and if a matrix A is non-singular then there exists a matrix A^{-1} called inverse of A such that

$A^{-1}A = AA^{-1} = I_{nxn}$ identity matrix of order n. Therefore if row-rank of $A_{nxn} = n$, then matrix is non-singular. If $\det A = 0$, then matrix A is not non-singular, and it is called singular.

We can determine, whether a given matrix A is singular or non-singular by reducing it to row echelon or row-reduced echelon form using elementary row-operations. If row-rank of A_{nxn} is less than n, then matrix is singular. If row-rank of $A_{nxn} = n$, then matrix is non-singular.

We reduce the given matrix A to echelon form or row reduced echelon form by applying elementary row operations to compute the determinant value of the matrix A, Record of interchange of any two rows and row devisors throughout elementary row operations is maintained. The determinant value of matrix $A = (-1)^r$ multiplication of diagonal entries of the reduced matrix $\times$ devisors, where r number of row interchanges

Definition 2.8: A matrix $(a_{ij})_{nxn}$ is called upper triangular if $a_{ij} = 0$ for $i > j$ and is called lower triangular if $a_{ij} = 0$ for $i < j$.

Remark: In case of upper triangular or lower triangular matrix, determinant value of the matrix = product of the diagonal entries.

System of Linear Equations

Let m be the row rank of a $n \times n$ square matrix A, obtained from row echelon form or row reduced echelon form by applying elementary row operations on A, If $m = n$ then matrix A is non-singular and if $m < n$, then A is a singular matrix and the determinant of A is zero. The following examples illustrate procedure of checking non-singularity of the matrix and evaluation of the determinant value of A.

Example 15: Using elementary row operations, check whether the following matrix is non-singular, if so then, find the determinant value of the matrix.

$$A = \begin{bmatrix} 3 & -2 & 4 & 5 \\ -1 & 3 & 2 & -2 \\ 2 & 4 & -5 & 3 \\ 4 & 5 & 1 & 6 \end{bmatrix}$$

$r_1 \leftarrow r_1 + 3r_2,\ r_3 \leftarrow r_3 + 2r_2,\ r_4 \leftarrow r_4 + 4r_2 \sim \begin{bmatrix} 0 & 7 & 10 & -1 \\ -1 & 3 & 2 & -2 \\ 0 & 10 & -1 & -1 \\ 0 & 17 & 9 & -2 \end{bmatrix}$

$r_1 \leftrightarrow r_2$ multiplying r_1 by -1 (record of sign changes two times)

$$\sim \begin{bmatrix} 1 & -3 & -2 & 2 \\ 0 & 7 & 10 & -1 \\ 0 & 10 & -1 & -1 \\ 0 & 17 & 9 & -2 \end{bmatrix}$$

$r_3 \leftarrow \dfrac{1}{10} r_3,\ r_3 \leftrightarrow r_2 \sim \begin{bmatrix} 1 & -3 & -2 & 2 \\ 0 & 1 & -0.1 & -0.1 \\ 0 & 7 & 10 & -1 \\ 0 & 17 & 9 & -2 \end{bmatrix}$

(Divided by 10 and sign changes 3 times),

$r_3 \leftarrow r_3 - 7r_2,\ r_4 \leftarrow r_4 - 17r_2 = \begin{bmatrix} 1 & -3 & -2 & 2 \\ 0 & 1 & -0.1 & -0.1 \\ 0 & 0 & 10.7 & -0.3 \\ 0 & 0 & 10.7 & -0.3 \end{bmatrix}_2$

$$r_4 \leftarrow r_4 - r_3 \sim \begin{bmatrix} 1 & -3 & -2 & 2 \\ 0 & 1 & -0.1 & -0.1 \\ 0 & 0 & 17.7 & -0.3 \\ 0 & 0 & 0 & 0 \end{bmatrix}$$

Number of nonzero rows in the above matrix is 3, less then the order of the matrix i.e. 4, therefore the given matrix is a singular matrix and multiplication of diagonal entries is zero, so the determinant value is zero.

Example 16: Using elementary row operations, check whether the given matrix is non-singular, if so find the determinant value of the matrix.

$$A = \begin{bmatrix} 3 & -2 & 4 & 5 \\ 1 & 2 & -2 & 3 \\ 2 & 4 & -5 & -2 \\ 4 & 0 & -3 & 2 \end{bmatrix}$$

$$r_1 \leftarrow r_1 - 3r_2,\ r_3 \leftarrow r_3 - 2r_2,\ r_4 \leftarrow r_4 - 4r_2 \sim \begin{bmatrix} 0 & -8 & 10 & -4 \\ 1 & 2 & -2 & 3 \\ 0 & 0 & -1 & -8 \\ 0 & -8 & 5 & -10 \end{bmatrix}$$

$$r_4 \leftarrow r_4 - r_1 \sim \begin{bmatrix} 0 & -8 & 10 & -4 \\ 1 & 2 & -2 & 3 \\ 0 & 0 & -1 & -8 \\ 0 & 0 & -5 & -6 \end{bmatrix}$$

$r_1 \leftrightarrow r_2$ (one-interchange), r_2 divided by -8, $-1 \times r_3$ multiplication of r_3, one sign change (sign changes 3 times, and divisor is 8)

$$\sim \begin{bmatrix} 1 & 2 & -2 & 3 \\ 0 & 1 & -1.25 & 0.5 \\ 0 & 0 & 1 & 8 \\ 0 & 0 & -5 & -6 \end{bmatrix}$$

$$r_4 \leftarrow r_4 + 5r_3 \sim \begin{bmatrix} 1 & 2 & -2 & 3 \\ 0 & 1 & 1.25 & 0.125 \\ 0 & 0 & 1 & 8 \\ 0 & 0 & 0 & 34 \end{bmatrix}$$

Now matrix is in echelon form, row-rank of matrix is 4 = order of the matrix, hence matrix is non-singular. Det $A = (-1) \times (-8) \times (-1) \times 34 = -272$.

Example 17: Reduce the following matrix in row-reduced echelon form and hence determine whether matrix is non-singular:

$$A = \begin{bmatrix} 2 & 0 & -1 & -3 \\ 1 & -2 & 0 & -1 \\ 1 & 2 & 4 & 1 \\ 3 & 2 & 3 & 5 \end{bmatrix}$$

on $r_1 \leftarrow r_1 - 2r_2$, $r_3 \leftarrow r_3 - r_2$, $r_4 \leftarrow r_4 - 3r_2$ we get $\sim \begin{bmatrix} 0 & 4 & -1 & -1 \\ 1 & -2 & 0 & -1 \\ 0 & 4 & 4 & 2 \\ 0 & 8 & 3 & 8 \end{bmatrix}$

on $r_1 \leftarrow r_1 - r_3$, $r_4 \leftarrow r_4 - 2r_3$, we get $\sim \begin{bmatrix} 0 & 0 & -5 & -3 \\ 1 & -2 & 0 & -1 \\ 0 & 4 & 4 & 2 \\ 0 & 0 & -5 & 4 \end{bmatrix}$

on $r_4 \leftarrow r_4 - r_1$ and $r_3 \leftarrow r_3/4$, we get $\sim \begin{bmatrix} 0 & 0 & -5 & -3 \\ 1 & -2 & 0 & -1 \\ 0 & 1 & 1 & 0.5 \\ 0 & 0 & 0 & 7 \end{bmatrix}$, (divisor is 4)

on $r_2 \leftarrow r_2 + 2r_3$ and $r_4 \leftarrow r_4/7$, we get $\sim \begin{bmatrix} 0 & 0 & -5 & -3 \\ 1 & 0 & 2 & 0 \\ 0 & 1 & 1 & 0.5 \\ 0 & 0 & 0 & 1 \end{bmatrix}$, (divisor is 7) on $r_1 \leftrightarrow r_2$ then $r_2 \leftrightarrow r_3$, we get $\sim \begin{bmatrix} 1 & 0 & 2 & 0 \\ 0 & 1 & 1 & 0.5 \\ 0 & 0 & -5 & -3 \\ 0 & 0 & 0 & 1 \end{bmatrix}$ (two sign changes)

on $r_2 \leftarrow r_2 - 0.5r_4$, $r_3 \leftarrow r_3 + 3r_4$, we get $\sim \begin{bmatrix} 1 & 0 & 2 & 0 \\ 0 & 1 & 1 & 0 \\ 0 & 0 & -5 & 0 \\ 0 & 0 & 0 & 1 \end{bmatrix}$

$r_3 \leftarrow r_3/-5$, $r_2 \leftarrow r_2 - r_3$ and $r_1 \leftarrow r_1 - 2r_3$, we get $\sim \begin{bmatrix} 1 & 0 & 0 & 0 \\ 0 & 1 & 0 & 0 \\ 0 & 0 & 1 & 0 \\ 0 & 0 & 0 & 1 \end{bmatrix}$, (divisor is -5)

Row rank of $A = 4 =$ order of matrix, hence matrix is non singular.

Det $A = (-)^2 \times 4 \times 7(-5) = -140$, since multiplication of diagonal entries is 1.

2.5 INVERSE OF A MATRIX

Inverse of A_{nxn} square matrix can be computed, using elementary row operations by bringing the matrix A to row-reduced echelon form.

Consider working matrix $W = [A, I]$, $n \times 2n$ matrix, where I is $n \times n$ identity matrix.

Now performing elementary row operations on W, such that A is brought to row-reduced echelon form and becomes I_{nxn} identity matrix in case of non-singular matrix A, then I_{nxn} would be converted to A^{-1} by the same sequence of elementary row operations done simultaneously.

Above argument is justified in the following lines. If A is non-singular, then A^{-1} exists. on applying A^{-1} on left of W, we get $A^{-1}W = A^{-1}[A, I] = [A^{-1}A, A^{-1}]$, since matrix multiplication is associative.

Further $A^{-1}W = [I, A^{-1}]$ i.e. matrix appearing on the place of I is inverse of A.

If A can not be brought to I in $[A, I]$ by elementary row operations, then it concludes that A is singular (row-rank of $A < n$), and inverse does not exist.

System of Linear Equations

Example 18: Find the inverse of matrix A by elementary row operations of the given

matrix $A = \begin{bmatrix} 3 & -2 & 1 \\ 1 & 3 & -2 \\ 2 & -1 & 3 \end{bmatrix}$.

Consider the working matrix $W = [A, I]$.

$$\sim \begin{bmatrix} 3 & -2 & 1 & 1 & 0 & 0 \\ 1 & 3 & -2 & 0 & 1 & 0 \\ 2 & 1 & 3 & 0 & 0 & 1 \end{bmatrix}$$

on $r_1 \leftrightarrow r_2$ $\sim \begin{bmatrix} 1 & 3 & -2 & 0 & 1 & 0 \\ 3 & -2 & 1 & 1 & 0 & 0 \\ 2 & -1 & 3 & 0 & 0 & 1 \end{bmatrix}$

on $r_2 \leftarrow r_2 - 3r_1, r_3 \leftarrow r_3 - 2r_1$ $\sim \begin{bmatrix} 1 & 3 & -2 & 0 & 1 & 0 \\ 0 & -11 & 7 & 1 & -3 & 0 \\ 0 & -7 & 7 & 0 & -2 & 1 \end{bmatrix}$

on $r_3 \leftarrow r_3/-7$ $\sim \begin{bmatrix} 1 & 3 & -2 & 0 & 1 & 0 \\ 0 & -11 & 7 & 1 & -3 & 0 \\ 0 & -1 & -1 & 0 & 0.2857 & -0.14286 \end{bmatrix}$

$r_2 \leftarrow r_2 + r_3$ and $r_1 \leftarrow r_1 - 3r_3$ $\sim \begin{bmatrix} 1 & 0 & -2 & 0 & 0.1429 & 0.42858 \\ 0 & 0 & -4 & 1 & 0.1427 & -1.57146 \\ 0 & 1 & -1 & 0 & 0.2857 & -0.14286 \end{bmatrix}$

on $r_3 \leftrightarrow r_2$ and then $r_3 \leftarrow r_3/-4$ $\sim \begin{bmatrix} 1 & 0 & 1 & 0 & 0.1429 & 0 \\ 0 & 1 & -1 & 0 & 0.2857 & 0 \\ 0 & 0 & 1 & -0.25 & -0.035675 & 0.392865 \end{bmatrix}$

on $r_1 \leftarrow r_1 - r_3$ and then $r_2 \leftarrow r_2 + r_3$ $\sim \begin{bmatrix} 1 & 0 & 0 & +0.25 & 0.178575 & 0.035715 \\ 0 & 1 & 0 & -0.25 & 0.250025 & 0.249999 \\ 0 & 0 & 1 & -0.25 & -0.035675 & 0.392865 \end{bmatrix}$

Therefore $A^{-1} = \begin{bmatrix} +0.25 & 0.178575 & 0.035715 \\ -0.25 & 0.250025 & 0.249999 \\ -0.25 & -0.035675 & 0.392865 \end{bmatrix}$.

In rational arithmetic $A^{-1} = \begin{bmatrix} \dfrac{1}{4} & \dfrac{5}{28} & \dfrac{1}{28} \\ -\dfrac{1}{4} & \dfrac{1}{4} & \dfrac{1}{4} \\ -\dfrac{1}{4} & -\dfrac{1}{28} & \dfrac{11}{28} \end{bmatrix}$.

Example 19: Using the row reduction method, check whether the given matrix A is invertible or not, if it is invertible, find A^{-1}.

$$A = \begin{bmatrix} -1 & 1 & 1 \\ 3 & 1 & -1 \\ 2 & 2 & 1 \end{bmatrix}$$

Working matrix $W = \begin{bmatrix} -1 & 1 & 1 & 1 & 0 & 0 \\ 3 & 1 & -1 & 0 & 1 & 0 \\ 2 & 2 & 1 & 0 & 0 & 1 \end{bmatrix}$

on $r_2 \leftarrow r_2 + 3r_1$, $r_3 \leftarrow r_3 + 2r_1$ ~ $\begin{bmatrix} -1 & 1 & 1 & 1 & 0 & 0 \\ 0 & 4 & 2 & 3 & 1 & 0 \\ 0 & 4 & 3 & 2 & 0 & 1 \end{bmatrix}$

on $r_3 \leftarrow r_3 - r_2$ and $r_1 \leftarrow r_1 \times -1$ ~ $\begin{bmatrix} 1 & -1 & -1 & -1 & 0 & 0 \\ 0 & 4 & 2 & 3 & 1 & 0 \\ 0 & 0 & 1 & -1 & -1 & 1 \end{bmatrix}$

on $r_2 \leftarrow r_2/4$ ~ $\begin{bmatrix} 1 & -1 & -1 & -1 & 0 & 0 \\ 0 & 1 & 0.5 & 0.75 & 0.25 & 0 \\ 0 & 0 & 1 & -1 & -1 & 1 \end{bmatrix}$

on $r_2 \leftarrow r_2 + r_1$ ~ $\begin{bmatrix} 1 & 0 & -0.5 & -0.25 & 0.125 & 0 \\ 0 & 1 & 0.5 & 0.75 & 0.25 & 0 \\ 0 & 0 & 1 & -1 & -1 & 1 \end{bmatrix}$

on $r_1 \leftarrow r_1 + 0.5r_3$ and $r_2 \leftarrow r_2 - 0.5r_3$ ~ $\begin{bmatrix} 1 & 0 & 0 & -0.75 & -0.25 & 0.5 \\ 0 & 1 & 0 & 1.25 & 0.75 & -0.5 \\ 0 & 0 & 1 & -1 & -1 & 1 \end{bmatrix}$

We know that row rank of $A = 3$, therefore matrix A is non-singular & inverse of matrix A is given by

$$A^{-1} = \begin{bmatrix} -0.75 & -0.25 & 0.5 \\ 1.25 & 0.75 & -0.5 \\ -1 & -1 & 1 \end{bmatrix}$$

Example 20: Using elementary row, operations, check whether the following matrix A is non-singular, if so find the inverse of A.

$$A = \begin{bmatrix} 3 & -2 & 1 & 4 \\ 1 & 3 & -2 & 5 \\ -2 & 1 & 5 & -3 \\ 2 & 2 & 4 & 6 \end{bmatrix}$$

Consider the working matrix $w = [A, I]$

$$\sim \begin{bmatrix} 3 & -2 & 1 & 4 & 1 & 0 & 0 & 0 \\ 1 & 3 & -2 & 5 & 0 & 1 & 0 & 0 \\ -2 & 1 & 5 & -3 & 0 & 0 & 1 & 0 \\ 2 & 2 & 4 & 6 & 0 & 0 & 0 & 1 \end{bmatrix}$$

On $r_1 \leftarrow r_1 - 3r_2$, $r_3 \leftarrow r_3 + 2r_2$, $r_4 \leftarrow r_4 - 2r_2$,

$$\sim \begin{bmatrix} 0 & -11 & 7 & -11 & 1 & -3 & 0 & 0 \\ 1 & 3 & -2 & 5 & 0 & 1 & 0 & 0 \\ 0 & 7 & 1 & 7 & 0 & 2 & 1 & 0 \\ 0 & -4 & 8 & -4 & 0 & -2 & 0 & 1 \end{bmatrix}$$

On $r_1 \leftrightarrow r_2$, $r_3 \leftrightarrow r_4$, then $r_2 \leftrightarrow r_3$

We get, ~ $\begin{bmatrix} 1 & 3 & -2 & 5 & 0 & 1 & 0 & 0 \\ 0 & -4 & 8 & -4 & 0 & -2 & 0 & 1 \\ 0 & -11 & 7 & -11 & 1 & -3 & 0 & 0 \\ 0 & 7 & 1 & 7 & 0 & 2 & 1 & 0 \end{bmatrix}$

2.24 Elementary Linear Algebra

On $r_2 \leftarrow r_2/-4$

$$\sim \begin{bmatrix} 1 & 3 & -2 & 5 & 0 & 1 & 0 & 0 \\ 0 & 1 & -2 & 1 & 0 & 0.5 & 0 & 0.25 \\ 0 & -11 & 7 & -11 & 1 & -3 & 0 & 0 \\ 0 & 7 & 1 & 7 & 0 & 2 & 1 & 0 \end{bmatrix}$$

On $r_3 \leftarrow r_3 + 11r_2,\ r_4 \leftarrow r_4 - 7r_2$

$$\sim \begin{bmatrix} 1 & 3 & -2 & 5 & 0 & 1 & 0 & 0 \\ 0 & 1 & -2 & 1 & 0 & 0.5 & 0 & 0.25 \\ 0 & 0 & -15 & 0 & 1 & 2.5 & 0 & 2.75 \\ 0 & 0 & 15 & 0 & 0 & -1.5 & 1 & -1.75 \end{bmatrix}$$

$$\sim \begin{bmatrix} 1 & 3 & -2 & 5 & 0 & 1 & 0 & 0 \\ 0 & 1 & -2 & 1 & 0 & 0.5 & 0 & 0.25 \\ 0 & 0 & -15 & 0 & 1 & 2.5 & 0 & 2.75 \\ 0 & 0 & 0 & 0 & 1 & 1 & 1 & 1 \end{bmatrix}$$

Last row of first half matrix is of zeros, therefore can not be brought to 0, 0, 0, 1 Hence, given matrix is singular, inverse of A does not exist.

Example 21: Find the inverse of the given matrix A, using elementary row operations, if it exists

$$A = \begin{bmatrix} 2 & 0 & 1 & -1 \\ 1 & -1 & 0 & 2 \\ 3 & 2 & 1 & 0 \\ 0 & 1 & 2 & 3 \end{bmatrix}$$

Working matrix

$$w = \begin{bmatrix} 2 & 0 & 1 & -1 & 1 & 0 & 0 & 0 \\ 1 & -1 & 0 & 2 & 0 & 1 & 0 & 0 \\ 3 & 2 & 1 & 0 & 0 & 0 & 1 & 0 \\ 0 & 1 & 2 & 3 & 0 & 0 & 0 & 1 \end{bmatrix}$$

$r_1 \leftarrow r_1 - 2r_2,\ r_3 \leftarrow r_3 - 3r_2$

System of Linear Equations

$$\begin{bmatrix} 0 & 2 & 1 & -5 & 1 & -2 & 0 & 0 \\ 1 & -1 & 0 & 2 & 0 & 1 & 0 & 0 \\ 0 & 5 & 1 & -6 & 0 & -3 & 1 & 0 \\ 0 & 1 & 2 & 3 & 0 & 0 & 0 & 1 \end{bmatrix}$$

$r_1 \leftrightarrow r_2$, than $r_2 \leftrightarrow r_4$

$$\sim \begin{bmatrix} 1 & -1 & 0 & 2 & 0 & 1 & 0 & 0 \\ 0 & 1 & 2 & 3 & 0 & 0 & 0 & 1 \\ 0 & 5 & 1 & -6 & 0 & -3 & 0 & 0 \\ 0 & 2 & 1 & -5 & 1 & -2 & 0 & 0 \end{bmatrix}$$

on $r_1 \leftarrow r_1 + r_2$, $r_3 \leftarrow r_3 - 5r_2$, $r_4 \leftarrow r_4 - 2r_2$

$$\sim \begin{bmatrix} 1 & 0 & 2 & 5 & 0 & 1 & 0 & 1 \\ 0 & 1 & 2 & 3 & 0 & 0 & 0 & 1 \\ 0 & 0 & -9 & -21 & 0 & -3 & 1 & -5 \\ 0 & 0 & -3 & -11 & 1 & -2 & 0 & -2 \end{bmatrix}$$

$r_3 \leftarrow -\dfrac{1}{9} r_3$

$$\sim \begin{bmatrix} 1 & 0 & 2 & 5 & 0 & 1 & 0 & 1 \\ 0 & 1 & 2 & 3 & 0 & 0 & 0 & 1 \\ 0 & 0 & 1 & 2.3333 & 0 & 0.33333 & -0.11111 & 0.55556 \\ 0 & 0 & -3 & -11 & 1 & -2 & 0 & -2 \end{bmatrix}$$

on $r_1 \leftarrow r_1 - 2r_3$, $r_2 \leftarrow r_2 - 2r_3$, $r_4 \leftarrow r_4 + 3r_3$

$$\sim \begin{bmatrix} 1 & 0 & 0 & 0.33333 & 0 & 0.33333 & 0.66667 & -0.1111 \\ 0 & 1 & 0 & -1.6667 & 0 & -0.66667 & 0.22222 & -0.1111 \\ 0 & 0 & 1 & 2.33333 & 0 & 0.33333 & -0.11111 & 0.55556 \\ 0 & 0 & 0 & -4 & 1 & -1 & -0.33333 & -0.33333 \end{bmatrix}$$

$r_4 \leftarrow -\dfrac{1}{4}r_4$

$$\sim \begin{bmatrix} 1 & 0 & 0 & 0.33333 & 0 & 0.33333 & 0.22222 & -0.11111 \\ 0 & 1 & 0 & -1.6667 & 0 & -0.66667 & 0.22222 & -0.11111 \\ 0 & 0 & 1 & 2.33333 & 0 & 0.33333 & -0.11111 & 0.55556 \\ 0 & 0 & 0 & 1 & -2.5 & 2.5 & 0.83333 & 0.83333 \end{bmatrix}$$

on $r_1 \leftarrow r_1 - \dfrac{1}{3}r_4, r_2 \leftarrow r_2 + \dfrac{5}{3}r_4, r_3 \leftarrow r_3 - \dfrac{7}{3}r_4$

$$\begin{bmatrix} 1 & 0 & 0 & 0 & 0.083333 & 0.25 & 0.19444 & 0.138889 \\ 0 & 1 & 0 & 0 & -0.41667 & -0.25 & 0.36111 & 0.02778 \\ 0 & 0 & 1 & 0 & 0.58333 & -0.25 & -0.30556 & 0.30556 \\ 0 & 0 & 0 & 1 & -0.25 & 0.25 & 0.083333 & 0.083333 \end{bmatrix}$$

Hence the inverse of the matrix is

$$A^{-1} = \begin{bmatrix} 0.083333 & 0.25 & 0.19444 & 0.138889 \\ -0.41667 & -0.25 & 0.36111 & 0.027778 \\ 0.58333 & -0.25 & -0.30556 & 0.30556 \\ -0.25 & 0.25 & 0.083333 & 0.083333 \end{bmatrix}.$$

Example 22: Find the inverse of the given matrix A, using elementary row operations, if it exists

$$A = \begin{bmatrix} 2 & -3 & 1 & 4 \\ 1 & 2 & 4 & -3 \\ -3 & 4 & 2 & 1 \\ 4 & 1 & -3 & 2 \end{bmatrix}$$

Working matrix $W = \begin{bmatrix} 2 & -3 & 1 & 4 & 1 & 0 & 0 & 0 \\ 1 & 2 & 4 & -3 & 0 & 1 & 0 & 0 \\ -3 & 4 & 2 & 1 & 0 & 0 & 1 & 0 \\ 4 & 1 & -3 & 2 & 0 & 0 & 0 & 1 \end{bmatrix}$

On $r_1 \leftarrow r_1 - 2r_2$, $r_3 \leftarrow r_3 + 3r_2$, $r_4 \leftarrow r_4 - 4r_2$

$$\sim \begin{bmatrix} 0 & -7 & -7 & 10 & 1 & -2 & 0 & 0 \\ 1 & 2 & 4 & -3 & 0 & 1 & 0 & 0 \\ 0 & 10 & 14 & -8 & 0 & 3 & 1 & 0 \\ 0 & -7 & -19 & 14 & 0 & -4 & 0 & 1 \end{bmatrix}$$

on $r_4 \leftarrow r_4 - r_1$ and $r_3 \leftarrow r_3/10$

$$\sim \begin{bmatrix} 0 & -7 & -7 & 10 & 1 & -2 & 0 & 0 \\ 1 & 2 & 4 & -3 & 0 & 1 & 0 & 0 \\ 0 & 1 & 1.4 & -0.8 & 0 & 0.3 & 0.1 & 0 \\ 0 & 0 & -12 & 4 & -1 & -2 & 0 & 1 \end{bmatrix}$$

on $r_2 \leftarrow r_2 - 2r_3$, $r_1 \leftarrow r_1 + 7r_3$ and $r_4 \leftarrow r_4/(-12)$

$$\sim \begin{bmatrix} 0 & 0 & 2.8 & 4.4 & 1 & 0.1 & 0.7 & 0 \\ 1 & 0 & 1.2 & -1.4 & 0 & 0.4 & -0.2 & 0 \\ 0 & 1 & 1.4 & -0.8 & 0 & 0.3 & 0.1 & 0 \\ 0 & 0 & 1 & -0.33333 & 0.083333 & 0.16667 & 0 & -0.083333 \end{bmatrix}$$

on $r_3 \leftarrow r_3 - 0.5\, r_1$, $r_2 \leftarrow r_2 - 1.2 r_4$

$$\sim \begin{bmatrix} 0 & 0 & 2.8 & 4.4 & 1 & 0.1 & 0.7 & 0 \\ 1 & 0 & 0 & -1 & -0.1 & 0.2 & -0.2 & 0.1 \\ 0 & 1 & 0 & -3 & -0.5 & 0.25 & -0.25 & 0. \\ 0 & 0 & 1 & -0.33333 & 0.083333 & 0.16667 & 0 & -0.083333 \end{bmatrix}$$

on $r_1 \leftarrow r_1 - 2.8 r_4$

$$\sim \begin{bmatrix} 0 & 0 & 0 & 5.3333 & 0.76667 & -0.36668 & 0.7 & 0.23333 \\ 1 & 0 & 0 & -1 & -0.1 & 0.2 & -0.2 & 0.1 \\ 0 & 1 & 0 & -3 & -0.5 & 0.25 & -0.25 & 0 \\ 0 & 0 & 1 & -0.33333 & 0.083333 & 0.16667 & 0 & -0.083333 \end{bmatrix}$$

on $r_1 \leftarrow r_1/5.3333$

$$\sim \begin{bmatrix} 0 & 0 & 0 & 1 & 0.14375 & -0.068753 & 0.13125 & 0.04375 \\ 1 & 0 & 0 & -1 & -0.1 & 0.2 & -0.2 & 0.1 \\ 0 & 1 & 0 & -3 & -0.5 & 0.25 & -0.25 & 0 \\ 0 & 0 & 1 & -0.33333 & 0.083333 & 0.16667 & 0 & -0.083333 \end{bmatrix}$$

on $r_2 \leftarrow r_2 + r_1,\ r_3 \leftarrow r_3 + 3r_1,\ r_4 \leftarrow r_4 + 0.33333 r_1$

$$\sim \begin{bmatrix} 0 & 0 & 0 & 1 & 0.14375 & -0.068753 & 0.13125 & 0.04375 \\ 1 & 0 & 0 & 0 & 0.04375 & 0.131247 & -0.06875 & 0.14375 \\ 0 & 1 & 0 & 0 & -0.06875 & 0.04374 & 0.14375 & 0.13125 \\ 0 & 0 & 1 & 0 & 0.13125 & 0.143753 & 0.04375 & -0.06875 \end{bmatrix}$$

on rearranging the rows

$$\sim \begin{bmatrix} 1 & 0 & 0 & 0 & 0.04375 & 0.131247 & -0.06875 & 0.14375 \\ 0 & 1 & 0 & 0 & -0.06875 & 0.04374 & 0.14375 & 0.13125 \\ 0 & 0 & 1 & 0 & 0.13125 & 0.143753 & 0.04375 & -0.06875 \\ 0 & 0 & 0 & 1 & 0.14375 & -0.068753 & 0.13125 & 0.04375 \end{bmatrix}$$

Now first half is identity matrix, therefore given matrix A is non-singular, hence

$$A^{-1} = \sim \begin{bmatrix} 0.04375 & 0.13125 & -0.06875 & 0.14375 \\ -0.06875 & 0.04375 & 0.14375 & 0.13125 \\ 0.13125 & 0.14375 & 0.04375 & -0.06875 \\ 0.14375 & -0.06875 & 0.13125 & 0.04375 \end{bmatrix}$$

EXERCISE SET 2

1. Using elementary row operations, check whether the following system of linear equations has solution. If it has solution whether solution is unique or many solutions. Find the solution(s) if exists.

$$3x_1 - 4x_2 + x_3 = -6$$
$$4x_1 - x_2 - 3x_3 = 5$$
$$x_1 + 5x_2 + 4x_3 = 7$$

System of Linear Equations

2. Using elementary row operations, check whether the following system of linear equations has solution. If it has solution whether solution is unique or many solutions. Find the solution(s) if exists.

$$3x_1 + 4x_2 + x_3 - x_4 = 5$$
$$4x_1 - x_2 + 3x_3 + 5x_4 = 2$$
$$x_1 + 5x_2 - 4x_3 - 6x_4 = 5$$
$$4x_1 + 4x_2 - x_2 = 6.$$

3. Using elementary row-operations, find the solution of the system of linear equations, if solution exists

$$2x_1 + x_2 - 3x_3 + x_4 = -4$$
$$3x_1 - 2x_2 + 4x_3 - 2x_4 = 18$$
$$5x_1 + 3x_2 + 3x_3 - x_4 = 6$$
$$4x_1 + 6x_2 - 4x_3 + 2x_4 = 6.$$

4. Using row reduced echelon form, check whether the following system is consistent if so, find the solution(s).

$$3x_1 + 4x_2 + 2x_3 = 4$$
$$4x_1 - 2x_2 + 3x_3 = 13$$
$$5x_1 + 3x_2 + 4x_3 = 11.$$

5. Using row reduced echelon form, check whether the following system is consistent if so, find the solution(s).

$$3x_1 + 4x_2 + x_3 - x_4 = 5$$
$$4x_1 - x_2 + 3x_3 + 5x_4 = 2$$
$$x_1 + 5x_2 - 4x_3 - 6x_4 = 5$$

6. Using row reduced echelon form, check whether the following system is consistent if so, find the solution(s).

$$3x_1 + 4x_2 + x_3 - x_4 = 14$$
$$4x_1 - x_2 + 3x_3 + 5x_4 = 9$$
$$x_1 + 5x_2 - 4x_3 - 6x_4 = 15$$
$$5x_1 - 3x_2 + 2x_3 + 2x_4 = 7.$$

7. Using row reduced echelon form, check whether the following system is consistent if so, find the solution(s).

$$3x_1 + 4x_2 + x_3 - x_4 = 3$$
$$4x_1 + x_2 + 3x_3 + 5x_4 = -1$$
$$x_1 - 5x_2 - 4x_3 + 6x_4 = -14$$
$$x_1 + x_2 + x_3 + x_4 = 0$$

8. Check whether the system of equations is consistent by reducing to row-echelon form

$$6x_1 + 3x_2 + x_3 = 12$$
$$x_1 + 5x_2 + 2x_3 = 3$$
$$2x_1 + 4x_2 + 7x_3 = 21$$
$$5x_1 + 4x_2 - 4x_3 = 6$$

9. Find the solution of the system of equations

$$6x_1 + 3x_2 + x_3 = 12$$
$$x_1 + 5x_2 + 2x_3 = 3$$
$$2x_1 + 4x_2 + 7x_3 = 21$$
$$5x_1 + 4x_2 - 4x_3 = -6$$

By reducing to row-echelon form, if solution exists.

10. Check by row-echelon form, whether following system of linear equations have solution, if so, find it:

$$x_1 + x_2 - 2x_3 + x_4 = -4$$
$$4x_1 - 2x_2 + x_3 + 2x_4 = 20$$
$$3x_1 - x_2 + 3x_3 - 2x_4 = 18$$
$$5x_1 - 3x_2 + 4x_3 - 3x_4 = 32$$

11. Find the solution of the following system of linear equation by reducing to row-echelon form

$$4x_1 + 3x_2 - 5x_3 = -20$$
$$3x_1 - 5x_2 + 4x_3 = 01$$
$$6x_1 + 7x_2 + 3x_3 = 04$$
$$7x_1 - 5x_2 + 9x_3 = 08$$

12. Find the solutions of the following system of linear equations by reducing to row-echelon form

$$4x_1 + 3x_2 - 5x_3 + 2x_4 = -12$$
$$3x_1 - 5x_2 + 4x_3 + 3x_4 = 13$$
$$6x_1 + 7x_2 + 3x_3 - 2x_4 = -4$$
$$7x_1 - 5x_2 + 9x_3 + x_4 = 12.$$

13. Check whether the following system of equation is consistent by reducing to row-reduced echelon form. If consistent, then find the solutions.

$$4x + 5y + 3z = -01$$
$$3x - 5y + 4z = 29$$

$$5x + y - 5z = -03$$
$$6x - 2y + 3z = 23.$$

14. Using row-reduced echelon form, check whether the following given system of linear equations is consistent. If so, find the solution.
$$x - y + z - w = -1$$
$$3x + 2y - 4z + 5w = -6$$
$$5x + 3y + z - 4w = 0$$
$$2x + 5y + 7z + 2w = 10.$$

15. Using row-reduced echelon form, check whether the system of linear equations is consistent. If so find the solution.
$$3x_1 + 2x_2 - x_3 = -1$$
$$-2x_1 + 3x_2 + 5x_3 = 6$$
$$5x_1 + x_2 - 2x_3 = 5$$
$$3x_1 + 3x_2 + x_3 = 3.$$

16. By reducing the following matrix into row reduced echelon form, find the value of the det A, where

$$A = \begin{bmatrix} 4 & 5 & -3 \\ 2 & -4 & 1 \\ 5 & 3 & 4 \end{bmatrix}$$

17. By reducing the following matrix into row reduced echelon form, find the value of the det A, where

$$A = \begin{bmatrix} 3 & 2 & -1 & -4 \\ 1 & 3 & 2 & 1 \\ -2 & 0 & 3 & 2 \\ 2 & 5 & 4 & -1 \end{bmatrix}.$$

18. By reducing the following matrix into row reduced echelon form, find the value of the det A, where

$$A = \begin{bmatrix} 3 & 2 & 0 & -4 \\ 1 & 3 & 2 & 0 \\ -2 & 0 & 3 & 2 \\ 2 & -5 & 4 & 1 \end{bmatrix}$$

19. Using row-reduced echelon form, check whether the given matrix A is invertible, if invertible, find A^{-1} by using elementary row operations.

$$A = \begin{bmatrix} 3 & 2 & 4 \\ 2 & 4 & 3 \\ 4 & -6 & 3 \end{bmatrix}.$$

20. Using row-reduction, check whether the given matrix A is invertible or not. Find A^{-1}, if it exists, where

$$A = \begin{bmatrix} -1 & 1 & 1 \\ 3 & 1 & -1 \\ 2 & 2 & 1 \end{bmatrix}.$$

21. Using row-reduction, check whether the given matrix A is invertible or not. Find A^{-1}, if it exists, where

$$A = \begin{bmatrix} 1 & 0 & 1 \\ 4 & -4 & 2 \\ 5 & 3 & 4 \end{bmatrix}.$$

22. By reducing the following matrix to row reduced echelon form, find the inverse of the matrix, where

$$A = \begin{bmatrix} 1 & 3 & 2 & 1 \\ -2 & 0 & 1 & 2 \\ 2 & -2 & -2 & 3 \\ 4 & 2 & 3 & -1 \end{bmatrix}.$$

23. Using elementary row operations, find the inverse of the given matrix A, if A is non-singular.

$$A = \begin{bmatrix} 2 & 1 & 3 & 3 \\ 3 & 5 & 4 & 6 \\ 5 & -5 & 4 & 2 \\ -4 & 9 & -3 & 1 \end{bmatrix}.$$

24. By reducing the following matrix into row reduced echelon form, check whether inverse exists, if so find the inverse

$$\begin{bmatrix} 3 & 2 & -1 & -4 \\ 1 & 3 & 2 & 1 \\ -2 & 0 & 3 & 2 \\ 2 & -5 & -4 & 1 \end{bmatrix}.$$

25. By reducing the following matrix to row reduced echelon form, find the value of the det A, where

$$A = \begin{bmatrix} 1 & 2 & -1 & -1 \\ 1 & 3 & 2 & 1 \\ -2 & 0 & 3 & 2 \\ 2 & -1 & -3 & 0 \end{bmatrix}.$$

26. Using row-reduced echelon form, find the inverse of the matrix A,

$$A = \begin{bmatrix} 3 & 2 & 2 & 0 \\ 2 & 4 & 2 & 0 \\ 3 & 2 & 3 & 0 \\ 2 & 4 & 1 & 1 \end{bmatrix}, \text{ if it exists.}$$

ANSWERS EXERCISE – 2

1. Unique $x_1 = 1$ $x_2, x_3 = -1$.   2. Infinite Solutions.   3. Solution does not exist
4. $x_1 = 2, x_2 = -1, x_3 = 1$, 5. Consistent, many Solutions. 6. $x_1 = 3, x_2 = 2, x_3 = -2, x_4 = 1$
7. $x_1 = 1, x_2 = -1, x_3 = 2, x_4 = -2$.   8. Inconsistent.   9. Consistent $x_1 = 2, x_2 = -1, x_3 = 3$
10. $x_1 = 3, x_2 = -3, x_3 = 2, x_4 = 0$. 11. $x_1 = -2, x_2 = 1, x_3 = 3$. 12. $x_1 = -2, x_2 = 1, x_3 = 3, x_4 = 4$. 13. Inconsistent. 14. $x = y = z = w = 1$. 15. Inconsistent
16. 17   17. 0,   18. 12

19. $\begin{bmatrix} -3 & 3 & 1 \\ -\frac{6}{10} & \frac{7}{10} & \frac{1}{10} \\ \frac{14}{5} & -\frac{13}{5} & -\frac{4}{5} \end{bmatrix}$. 20. $\begin{bmatrix} -\frac{3}{4} & -\frac{1}{4} & \frac{1}{2} \\ \frac{5}{4} & \frac{3}{4} & -\frac{1}{2} \\ -1 & -1 & 1 \end{bmatrix}$. 21. $\begin{bmatrix} -\frac{22}{10} & \frac{3}{10} & \frac{4}{10} \\ -\frac{6}{10} & -\frac{1}{10} & \frac{2}{10} \\ \frac{32}{10} & -\frac{3}{10} & -\frac{4}{10} \end{bmatrix}$

22. $\begin{bmatrix} \frac{6}{10} & -\frac{4}{10} & \frac{2}{10} & 0 \\ -\frac{1}{10} & \frac{9}{10} & -\frac{2}{10} & -\frac{5}{10} \\ \frac{6}{10} & \frac{6}{10} & \frac{2}{10} & -\frac{10}{10} \\ -\frac{3}{10} & -\frac{13}{10} & \frac{4}{10} & \frac{15}{10} \end{bmatrix}$. 23. $\begin{bmatrix} -\frac{17}{20} & \frac{37}{20} & -\frac{14}{20} & -\frac{1}{20} \\ \frac{14}{20} & -\frac{14}{20} & \frac{8}{20} & \frac{2}{20} \\ -\frac{16}{20} & \frac{16}{20} & -\frac{12}{20} & -\frac{8}{20} \\ \frac{7}{20} & -\frac{7}{20} & \frac{14}{20} & \frac{11}{20} \end{bmatrix}$

24. Singular matrix, inverse does not exist.
25. Singlular matrix, inverse does not exist

26. $\begin{bmatrix} 1 & -\frac{1}{4} & -\frac{1}{2} & 0 \\ 0 & \frac{3}{8} & -\frac{1}{4} & 0 \\ -1 & 0 & 1 & 0 \\ -1 & -1 & 1 & 1 \end{bmatrix}$

CHAPTER 3

Vector Spaces

INTRODUCTION

A vector space (or linear space) is the basic object of study in the branch of mathematics called linear algebra. If one considers geometrical vectors and the operations, one can perform upon these vectors such as addition of vectors, scalar multiplication, with some natural constraints such as closure of these operations, associatively of these and combinations of these operations, and so on, we arrive at a description of a mathematical structure which we call a vector space. The "vectors" need not be geometric vectors in the normal sense, but can be any mathematical object that satisfies the following vector space axioms. Polynomials of degree $\leq n$ with real-valued coefficients form a vector space, for example. It is this abstract quality that makes it useful in many areas of modern mathematics. We discuss some examples for the better understanding of vector space.

3.1 VECTOR SPACES

Some examples are being discussed before giving formal definition of the vector space. For discussion of set of polynomials, we need the formal definition of degree of a polynomial

Definition 3.1: A polynomial denoted by $p_n(x) = a_0 + a_1 x + \ldots + a_{n-1} x^{n-1} + a_n x^n$ is called of degree n if $a_n \neq 0$. From this any non-zero real number is a polynomial of degree 0. Though for real number 0 this does not remain valid, because $a_0 = 0$, yet for the purpose of set of polynomials of degree $\leq n$, 0 is also included as a polynomial of degree 0.

Example 1: Consider $V = \{p(x) = a_0 x^2 + a_1 x + a_2 \mid a_0, a_1, a_2 \text{ are any real numbers}\}$ set of all polynomials of degree ≤ 2. Addition of two polynomials and multiplication of a polynomial by a real number is taken as usual and the following properties are noted.

3.2 Elementary Linear Algebra

Let $p(x), q(x) \in V$, where $p(x) = a_0x^2 + a_1x + a_2$, $q(x) = b_0x^2 + b_1x + b_2$ and $a_0, a_1, a_3, b_0, b_1, b_2$ be any real numbers

Addition of two polynomials is defined as

(a) $(p+q)(x) = p(x) + q(x) = a_0x^2 + a_1x + a_2 + b_0x^2 + b_1x + b_2$
$= (a_0 + b_0)x^2 + (a_1 + b_1)x + (a_2 + b_2)$

This polynomial is again in V, since is of degree ≤ 2.

i.e. $(p+q)(x) \in V$ this property of set is called the set V is closed under additions.

(b) $(p+q)(x) = (a_0 + b_0)x^2 + (a_1 + b_1)x + (a_2 + b_2)$
$= (b_0 + a_0)x^2 + (b_1 + a_1)x + (b_2 + a_2)$
$= (q+p)(x)$ addition of polynomials is commutative,

since sum of real numbers a_0, b_0, a_1, b_1, a_2 and b_2 is commutative.

(c) With $r(x) = c_0x^2 + c_1x + c_2$, c_0, c_1, c_2 are any real number

$(p+q)(x) + r(x) = (a_0 + b_0)x^2 + (a_1 + b_1)x + (a_2 + b_2) + c_0x^2 + c_1x + c_2$
$= (a_0 + b_0 + c_0)x^2 + (a_1 + b_1 + c_1)x + (a_2 + b_2 + c_2)$
$= a_0x^2 + a_1x + a_2 + (b_0 + c_0)x^2 + (b_1 + c_1)x + (b_2 + c_2)$
$= p(x) + (q+r)(x)$

Addition of polynomials is associative.

(d) Zero polynomial $0(x) = 0x^2 + 0x + 0 = 0 \in V$ and

$(p+0)x = a_0x^2 + a_1x + a_2 + 0 = 0 + a_0x^2 + a_1x + a_2$
$= (0+p)x = p(x)$. Zero polynomial exists in V.

(e) For each $p(x)$ in V, there is $-p(x) = -a_0x^2 - a_1x - a_2 \in V$, such that

$p(x) + (-p(x)) = (a_0 - a_0)x^2 + (a_1 - a_1)x + (a_2 - b_2)$
$= 0 = -p(x) + p(x)$

Existence of $-p(x)$ additive inverse, of $p(x)$ is in V.

(f) Multiplication of polynomial $p(x) = a_0x^2 + a_1x + a_2$ by real number α is defined as $(\alpha p)(x) = \alpha a_0x^2 + \alpha a_1x + \alpha a_2 = \alpha p(x) \in V$, this multiplication is known as scalar multiplication of $p(x)$. This property is known as V is closed under scalar multiplication.

This satisfies the following properties (g) to (j).

(g) $$\alpha(p(x) + q(x)) = \alpha(a_0 + b_0)x^2 + \alpha(a_1 + b_1)x + \alpha(a_2 + b_2)$$
$$= \alpha a_0 x^2 + \alpha a_1 x + \alpha a_2 + \alpha b_0 x^2 + \alpha b_1 x + \alpha b_2$$
$$= \alpha p(x) + \alpha q(x)$$

Scalar multiplication is distributive over addition of polynomials.

(h) $$(\alpha + \beta)(p(x)) = (\alpha + \beta)a_0 x^2 + (\alpha + \beta)a_1 x + (\alpha + \beta)a_2$$
$$= \alpha a_0 x^2 + \alpha a_1 x + \alpha a_2 + \beta a_0 x^2 + \beta a_1 x + \beta a_2$$
$$= \alpha(p(x)) + \beta q(x),$$

for any scalars α and β.

(i) $$\alpha(\beta(p(x))) = \alpha(\beta a_0 x^2 + \beta a_1 x + \beta a_2) = \alpha\beta a_0 x^2 + \alpha\beta a_1 x + \alpha\beta a_2$$
$$= (\alpha\beta)(p(x)) = \beta(\alpha(p(x))).$$

Scalar multiplication is commutative.

(j) $1p(x) = p(x)$, existence of unit scalar.

We have noted that certain conditions are satisfied in the example.

Example 2: Consider the set $V = \{u = (x_1, x_2, ..., x_n) | x_i \text{ are real numbers}\}$ of n-tuples of real numbers.

Addition of $u = (x_1, x_2, ..., x_n)$ and $v = (y_1, y_2, ..., y_n)$ of elements of V is defined as $u + v = (x_1 + y_1, x_2 + y_2, ..., x_n + y_n)$, and multiplication of u by scalar α is defined as $\alpha u = (\alpha x_1, \alpha x_2, ..., \alpha x_n)$.

We note the following properties in this example.

(a) $u + v \in V$, since $x_i + y_i$ is again a real number for $i = 1, 2, ..., n$

This shows that set V is closed under addition defined as above.

(b) $$u + v = (x_1 + y_1, x_2 + y_2, ..., x_n + y_n)$$
$$= (y_1 + x_1, y_2 + x_2, ..., y_n + x_n),$$

since addition of real numbers is commutative

$$= (y_1, y_2, ..., y_n) + (x_1, x_2, ..., x_n)$$
$$= v + u, \text{ addition is commutative.}$$

(c) $$(u + v) + w = (x_1 + y_1, x_2 + y_2, ..., x_n + y_n) + (z_1, z_2, ..., z_n)$$

where $w = (z_1, z_2, ..., z_n)$

$$= (x_1 + y_1 + z_1, x_2 + y_2 + z_2, ..., x_n + y_n)$$
$$= (x_1, x_2, ..., x_n) + (y_1 + z_1, y_2 + z_2, ..., y_n + z_n)$$
$$= u + (v + w), \text{ Addition is associative.}$$

(d) $\mathbf{0} = (0, 0, ..., 0)$, n zeros is defined as $\mathbf{0}$ element, satisfies

$$u + \mathbf{0} = (x_1 + 0, x_2 + 0, ..., x_n + 0) = (x_1, x_2, ..., x_n) = u = \mathbf{0} + u$$

This shows existence of $\mathbf{0} \in V$ zero element.

(e) Negative of u is defined as $-u = (-x_1, -x_2, ..., -x_n) \in V$, since $-x_1, -x_2, ..., -x_n \in R$.

$$u + (-u) = (x_1 - x_1, x_2 - x_2, ..., x_n - x_n) = (0, 0, ..., 0) = \mathbf{0} = (-u) + u \text{ Existence of } -u$$, additive inverse of u.

Multiplication of $u = (x_1, ..., x_n)$ by scalar α defined as $\alpha u = (\alpha x_1, \alpha x_2, ..., \alpha x_n)$, satisfies the conditions from (f) to (i).

(f) $\alpha(u + v) = \alpha((x_1, x_2, ..., x_n) + (y_1, ..., y_n))$
$$= \alpha(x_1, y_1, x_2 + y_2, ..., x_n + y_n), \text{ by addition}$$
$$= (\alpha(x_1 + y_1), \alpha(x_2 + y_2), ..., \alpha(x_n + y_n)), \text{ by scalar multiplication}$$
$$= (\alpha x_1 + \alpha y_1, \alpha x_2 + \alpha y_2, ..., \alpha x_n + \alpha y_n)$$
$$= (\alpha x_1, \alpha x_2, ..., \alpha x_n) + (\alpha y_1, \alpha y_2, ..., \alpha y_n)$$
$$= \alpha(x_1, x_2, ... x_n) + \alpha(y_1, y_2, ..., y_n)$$
$$= \alpha u + \alpha v, \text{ scalar multiplication is distributive over addition.}$$

(g) $(\alpha + \beta)u = ((\alpha + \beta)x_1, (\alpha + \beta)x_2, ..., (\alpha + \beta)x_n)$
$$= (\alpha x_1 + \beta x_1, \alpha x_2 + \beta x_2, ..., \alpha x_n + \beta x_n)$$
$$= (\alpha x_1, \alpha x_2, ..., \alpha x_n) + (\beta x_1, \beta x_2, ..., \beta x_n)$$
$$= \alpha(x_1, x_2, ..., x_n) + \beta(x_1, x_2, ..., x_n)$$
$$= \alpha u + \beta u.$$

(h) Similarly $\beta(\alpha u) = (\beta \alpha)u = \alpha(\beta u)$ scalar multiplication is associative.

(i) $1u = u$, existence of unit scalar 1.

In this example also some conditions are satisfied.

Vector Spaces

Example 3: Let $c[a, b]$ be the set all of continuous real valued functions on the interval $[a, b]$.

Point-wise addition of functions $f, g \in C[a, b]$ is defined as $(f + g)(x) = f(x) + g(x)$ for each $x \in [a, b]$ for a given $x \in [a, b]$, since $f(x)$ and $g(x)$ are real numbers, which are added, resulting again in a real number, This real number is the value of function $(f + g)$ at x, this should hold for each $x \in [a, b]$.

(a) We know that sum of continuous functions is continuous, therefore $(f + g)$ is continuous, hence belongs to $C[a, b]$, i.e. $C[a, b]$ is closed under addition defined as above. Further we not:

(b) $(f + g)(x) = f(x) + g(x) = g(x) + f(x) = (g + f)(x)$, for each $x \in [a, b]$, since addition of real numbers is commutative, therefore addition of functions is commutative.

(c) $((f + g) + h)(x) = (f + g)(x) + h(x) = f(x) + g(x) + h(x)$
$= f(x) + (g + h)(x) = (f + (g + h))(x)$,

for each $x \in [a, b]$ point-wise addition of continuous functions is associative.

(d) Zero function defined as $0(x) = 0$ for all $x \in [a, b]$, satisfies
$(f + 0)(x) = f(x) + 0(x) = f(x) + 0 = f(x) = 0 + f(x)$
$= 0(x) + f(x) = (0 + f)(x)$ for each $x \in [a, b]$

i.e. $\quad f + 0 = f = 0 + f$. Existence of zero function in $C[a, b]$.

(e) $(-f)(x) = -f(x)$ is defined for each $x \in [a, b]$ and for each $f \in C[a, b]$, satisfies
$(f + (-f))x = f(x) + (-f)(x) = f(x) - f(x) = 0$
$= -f(x) + f(x) = ((-f) + f)x$

i.e. $\quad f + (-f) = 0 = (-f) + (f)$ for each $f \in C[a, b]$

$-f$ existence of additive inverse of f.

(f) Multiplication of continuous function $f(x)$ by real number defined as $(\alpha f)(x) = \alpha f(x)$, for each $x \in [a, b]$, satisfies conditions from (g) to (i).

(g) $(\alpha(f + g))(x) = \alpha(f(x) + g(x)) = \alpha f(x) + \alpha g(x)$
$= (\alpha f)(x) + (\alpha g)(x)$, for each $x \in [a, b]$

i.e. $= (\alpha f)(x) + (\alpha g)(x)$, scalar multiplication is distributive over point-wise addition of functions for all $f, g \in C[a, b]$.

(h) $((\alpha + \beta)f)(x) = (\alpha + \beta)f(x) = \alpha f(x) + \beta f(x) = (\alpha f)(x) + (\beta f)(x)$
$= (\alpha f + \beta f)(x)$, for each $x \in [a, b]$

i.e. $(\alpha + \beta)f = \alpha f + \beta f$ for each $f \in c[a, b]$,
addition of scalar is distributive over multiplication of function.

(i) Similarly $\alpha(\beta f)(x) = (\alpha\beta) f(x) = \beta(\alpha f(x)) = \beta(\alpha f)(x))$

for each $x \in [a, b]$ and for all $f \in C[a, b]$, scalar multiplication is associative.

(j) $(1f)(x) = f(x)$ for each $x \in [a, b]$ and for all $f \in C[a, b]$, existence of unit scalar.

We note that three examples discussed above have some common properties, there are many other such sets having these properties.

Now we identify these properties and assign a name of sets with such properties by the following formal definition.

Definition 3.2: **Vector Space:** Let a non-empty set V of elements be given and a rule of addition of any two elements of V, denoted by + and another rule of multiplication of elements of V by real numbers (or complex numbers) also be defined, then set V is called vector space and elements of V are called vectors of V and real numbers (or complex numbers) are known as scalars, provided the following conditions are satisfied:

1. For any $u, v \in V$, addition + of u, v is defined and $u + v \in V$, this property is known as set is closed under addition, further

   (a) $u + v = v + u \quad \forall\ v \in V$, addition is commutative

   (b) $(u + v) + w = (u + v) + w \quad \forall u, v, w \in V$, addition is associative

   (c) There is an element called zero element denoted by 0_V in V, such that $u + 0_V = 0_V, u + u$ for each $u \in V$, existence of additive identity

   (d) For each $u \in V$, there exists $-u$, called additive inverse of u, such that $(-u) + u = u + (-u) = 0_V$, existence of additive inverse

2. Multiplication of vectors by a scalar (real or complex numbers), satisfies $\alpha u \in V$ for each $u \in V$ and each scalar α, the set V is called closed under scalar multiplication, and further

   (e) $\alpha(u + v) = \alpha u + \alpha v \quad \forall\ u, v \in V$, α, any scalar

   (f) $(\alpha + \beta)u = \alpha u + \beta u \quad \forall\ u \in V, \alpha, \beta$, any scalars

   (g) $\alpha(\beta u) = (\alpha\beta)u = \beta(\alpha u), \quad \forall\ u \in V, \alpha, \beta$ are any scalar multiplication is associative.

   (h) $1\ u = u, \forall\ u \in V$, **1** is known as unit scalar in (real or complex) number.

In example 1, Set V of polynomials of degree ≤ 2 now denoted by P_2, in example 2 set V of n-tuples denoted by V_n or R^n, and $C[a, b]$ = set of all continuous functions on interval $I = [a, b]$ satisfy all the above conditions. Therefore P_2, R^n and $C[a, b]$ are vector spaces.

For clarification of the concept of vector space, some more examples are given.

Vector Spaces

Example 4: Set $M_{m \times n}$ of all $m \times n$ matrices is a vector space with usual addition of matrices as vector addition and multiplication of a matrix by a scalar as scalar multiplication.

Addition of $m \times n$ matrices give an $m \times n$ matrix and multiplication of an $m \times n$ matrix by a scalar again yields an $m \times n$ matrix, therefore $M_{m \times n}$ is closed under vector addition and scalar multiplication.

An $M_{m \times n}$ matrix with all entries zeroes is the usual zero matrix. For any matrix $M_{m \times n}$, $-A$ is to be the matrix $(-1)A$.

All other conditions of vector space are satisfied. Therefore $M_{m \times n}$ is a vector space.

Example 5: Let F be the set of all real-valued functions mapping R into R.

Let $f, g \in F$. The vector sum of the functions be defined by

$$(f + g)(x) = f(x) + g(x) \text{ for all real numbers } x$$

For any scalar a in R and function f, the αf is the function, whose value at x is $\alpha f(x)$, so that $(\alpha f)(x) = \alpha f(x)$ for all $x \in R$.

Now to show that set F with these operations is a vector space.

We observe that, for f and g in F, both $f + g$ and af are functions mapping R into R, therefore $f + g$ and αf are in F. Thus, F is closed under vector addition and under scalar multiplication.

The constant function, whose value at each x in R is 0 is taken as zero function in F. For each function f in F, $-f$ that is the function $(-1)f$ in F is taken as additive inverse of f. Since the function $f + (-f) = f + (-1)f$ has as its value at x in R the number $f(x) + (-1)f(x) = f(x) - f(x) = 0$, for all $x \in R$. Consequently, $f + (-f)$ is the zero function from R to R. Similarly other conditions can be verified easily.

The scalar multiplicative properties are also easy to verify. For example, to verify 4th condition, we must compute $1(f)$ at any x in R and compare the result with $f(x)$. We obtain $(1f)(x) = 1f(x) = f(x)$ same value, therefore $1(f) = f$.

Hence F is a vector space.

Example 6: Let P_n be the set of all polynomials of degree $\leq n$, Consider the Set $V = \{p(x) \in P_n(x)\, p(1) = 0\}$, with addition of polynomials and multiplication by scalars is same as on $P_n(x)$.

For $p, q \in V$, $p(1) = 0$, $q(1) = 0$, i.e. $(p + q)(1) = p(1) + q(1) = 0 + 0 = 0 \Rightarrow p + q \in V$.

$(p + q)(x) = (q + p)(x) \Rightarrow (p + q)(1) = (q + p)(1) = 0 \Rightarrow$ For $p + q = q + r$.

$((p + q) + r)(x) = (p + (q + r))(x) \Rightarrow ((p + q) + r)(1) = (p + (q + r))(1)$

$\Rightarrow (p + q) + r = p + (q + r)$, for p, q and $r \in V$.

For all $p \in V$, $(-p + p)(x) = -p(x) + p(x) = 0$,
therefore $(-p + p)(1)(p + (-p))$ $(1) = 0$, $(-p)$ additive inverse of p exists.

Similarly all other conditions of scalar product are also satisfied.

Hence set V is a vector space.

Example 7: Set $V = \{(x_1, 0, x_3, x_4, x_5) | x_1, x_3, x_4, x_5$ are any real numbers$\}$ is a vector space. Addition and scalar multiplication are same as in R^n.

All conditions of vector space of R^5 are true for the given set except two conditions. These two conditions are shown below:

Since, for $\quad u = (x_1, 0, x_3, x_4, x_5)$, $v = (y_1, 0, y_3, y_4, y_5) \in V$.

$\quad\quad u + v = (x_1 + y_1, 0, x_3 + y_3, x_4 + y_4, x_5 + y_5) \in V$

And $\quad\quad \alpha u = \alpha(x_1, 0, x_3, x_4, x_5) = (\alpha x_1, 0, \alpha x_3, \alpha x_4, \alpha x_5) \in V$.

Therefore V is a vector space.

Example 8: $V = \left\{ f \in C[a,b] \bigg| f\left(\dfrac{a+b}{2}\right) = 0 \right\}$, is a vector space, with same addition of functions and scalar multiplication as defined in example 3.

Let $\quad f(x), g(x) \in V, i.e., f\left(\dfrac{a+b}{2}\right) = 0$ and $g\left(\dfrac{a+b}{2}\right) = 0$

$$(f+g)\left(\dfrac{a+b}{2}\right) = f\left(\dfrac{a+b}{2}\right) + g\left(\dfrac{a+b}{2}\right) = 0 + 0 = 0.$$

Therefore $f + g \in V$.

Also $(f + g) + h = f + (g + h)$ and $(f + g) = (g + f)$ are true since $f, g, h \in C[a, b]$.

$O\left(\dfrac{a+b}{2}\right) = 0$, defines zero vector of V.

For $f \in V$, $-f \in V$, since $(f+g)\left(\dfrac{a+b}{2}\right) = p\left(\dfrac{a+b}{2}\right) + g\left(\dfrac{a+b}{2}\right) = 0 + 0 = 0 - f\left(\dfrac{a+b}{2}\right)$ $= 0$.

Other conditions of scalar multiplications are true due to vectors of $C[a, b]$.

Hence V is a vector space.

Proposition: In any vector space V the following statements are true

(a) $\alpha 0_V = 0_V$, for any scalar α, where 0_V is zero vector of V.

(b) $0u = 0_V$, for each $u \in V$.

(c) $-1u = -u$, for each $u \in V$.

These are some simple facts of vectors, which can be proved easily.

The following sets are not vector spaces.

Example 9: Set of all polynomials of degree n, with usual rule of addition and scalar multiplication is not a vector space, since zero polynomial is not present in the set.

Example 10: Set of all continuous functions $f(x)$ on $[a, b]$, such that such that $f\left(\dfrac{a+b}{2}\right) = 1$, with usual addition of functions and multiplication by real numbers is not a vector space, since, $0\left(\dfrac{a+b}{2}\right) = 0$. Zero function is not in the set.

3.2 SUBSPACE

If V is a vector space over the real numbers (or complex), there are certain subsets of V, called subspaces, which are again vector spaces with the same operations of addition of elements and scalar multiplication. The purpose of this section is to study such subsets.

Definition 2.3: A non-empty subset S of a vector space V is called a subspace of V, if S itself is a vector space with rule of addition and scalar multiplication same as on vectors of V.

(**Note:** If S is a subset of V, but rule of addition or rule of scalar multiplication or both are not same as for V, then S is not a subspace of V).

Example 11: Let $V_5 = \{(x_1, x_2, ..., x_5)\ x_i\text{'s real numbers}\}$ be the vector space with addition and scalar multiplication as defined earlier.

A set $S = \{(x_1, x_2, ..., x_5) \in V_5 | a_1 x_1 + a_2 x_2 + a_3 x_3 + a_4 x_4 + a_5 x_5 = 0\}$, for some fixed numbers a_1, a_2, a_3, a_4 and a_5, is a subspace of V_5.

To show that S is a subspace of V, we should check all the conditions of a vector space on S also. Vectors of S are taken from vector space V, therefore checking all the conditions may not be necessary. Some conditions may be taken for granted. Therefore, we have a important result to reduce the work of checking of all the axioms of a vector space.

Theorem 3.1: A non-empty subset S of a vector space V is a subspace of V, if and only if (iff to be used for if and only if).

(a) If $u, v \in S$, then $u + v \in S$ for all $u, v \in S$.

(b) If $u \in S$, then $\alpha u \in S$ for each $u \in S$ and every scalar α.

Proof: If (a) and (b) are true, then $u + v = v + u$ since $u, v \in V$

$$(u + v) + w = u + (v + w) \text{ since } u, v, w \in V$$

Choosing $\alpha = -1$ in condition (b), $-1u = -u \in S$, additive inverse of u is in S, choosing $a = 0$ in (b), $0u = 0_V \in S$, existence of zero vector in S.

Therefore all conditions of vector additions are satisfied.

All the conditions of scalar multiplication are satisfied, because all vectors of S are vectors of V. Conversely, if S is a subspace of V, then S is a vector space in itself, therefore

for any $u, v \in S$, $u + v \in S$ $\forall\ u, v \in S$

and for any $u \in S$ $\alpha u \in S$ for any scalar α.

Note: Above conditions are also equivalent to a single condition that S is a subspace of V iff $\alpha u + \beta v \in S$, for any scalars α, β and for all $u, v \in S$.

Remark: Smallest subspace of any vector space is space of zero vector only, which is denoted by $V_0 = \{0_V\}$ and is called trivial vector space.

Now to check whether set in **example 11** is a subspace or not, we have to check only two conditions

First to show that S is non-empty set because $(0, 0, 0, 0, 0) \in S$ satisfies the conditions, therefore at least $0_V \in S$ i.e., S is non-empty set.

Now Let $u = (x_1, x_2, ..., x_n)$ and $v = (y_1, y_2, ..., y_n)$ be in S.

Therefore
$$a_1 x_1 + a_2 x + , ..., + a_n x_n = 0 \quad \text{(A)}$$

$$a_1 y_1 + a_2 y_2 + , ..., + a_n y_n = 0 \quad \text{(B)}$$

On adding above two relations (A) and (B), we have

$$a_1(x_1 + y_1) + a_2(x_2 + y_2) + , ..., + a_n(x_n + y_n) = 0.$$

This shows that $(x_1 + y_1, x_2 + y_2, x_3 + y_3, ..., x_n + y_n) \in S$

i.e. $(x_1, x_2, ..., x_n) + (y_1, y_2, ..., y_n) \in S \Rightarrow u + v \in S$.

From (A)
$$\alpha(a_1 x_1 + a_2 x_2 + , ..., + a_n x_n) = 0,$$

$$\alpha a_1 x_1 + \alpha a_2 x_2 + \alpha a_3 x_3 + , ..., + \alpha a_n x_n = 0,$$

$$a_1(\alpha x_1) + a_2(\alpha x_2) + , ..., + a_n(\alpha x_n) = 0,$$

Therefore $(\alpha x_1, \alpha x_2, ..., \alpha x_n) \in S \Rightarrow \alpha(x_1, x_2, ..., x_n) \in S$

i.e., $\alpha u \in S$ for all $u \in S$ and α is any scalar.

Hence S is a subspace of V.

Example 12: Let $V = R_3$ be the vector space as usual.

Let $U = \{u = (x_1, x_2, x_3) \in V \mid x_1 + x_2 + x_3 = 0\}$. Now to show that U is a subspace of V.

Let $u = (x_1, x_2, x_3)$ and $v = (y_1, y_2, y_3) \in U$, then $x_1 + x_2 + x_3 = 0$, and $y_1 + y_2 + y_3 = 0$ by definition of U.

$$u + v = (x_1 + y_2, x_2 + y_2, x_3 + y_3).$$

Now $(x_1 + y_1) + (x_2 + y_2) + (x_3 + y_3) = (x_1 + x_2 + x_3) + (y_1 + y_2 + y_3) = 0 + 0 = 0$ from above, This shows that $u + v \in U$.

Now $u = \alpha(x_1, x_2, x_3) = (\alpha x_1, \alpha x_2, \alpha x_3)$, $\alpha x_1 + \alpha x_2 + \alpha x_3 = \alpha(x_1 + x_2 + x_3) = \alpha 0 = 0$. This shows $\alpha u \in u$.

Therefore using above theorem, U is a subspace of $V = R^3$

Example 13: Let $V_3 = \{(x_1, x_2, x_3)\}$ with usual addition and scalar multiplication.

Let U be a subset of V_3 defined by $U = \{(x_1, x_2, x_3) \in V_3 \mid x_1 x_2 + x_2 x_3 = 0\}$

Let $u, v \in U$, $u = (x_1, x_2, x_3)$, $v = (y_1, y_2, y_3)$, then, $x_1 x_2 + x_2 x_3 = 0$, $y_1 y_2 + y_2 y_3 = 0$

Now consider $u + v = (x_1 + y_1, x_2 + y_2, x_3 + y_3)$.

Now to check $(x_1 + y_1)(x_2 + y_2) + (x_2 + y_2)(x_3 + y_3) = 0$.

$(x_1 + y_1)(x_2 + y_2) + (x_2 + y_2)(x_3 + y_3) = (x_2 + y_2)(x_1 + y_1 + x_3 + y_3)$

$= (x_2 + y_2)(x_1 + x_3 + y_1 + y_3) = (x_2 + y_2)(x_1 + x_3) + (x_2 + y_2)(y_1 + y_3)$

$= x_2(x_1 + x_3) + y_2(x_1 + x_3) + x_2(y_1 + y_3) + y_2(y_1 + y_3)$

$= 0 + y_2(x_1 + x_3) + x_2(y_1 + y_3) + 0$

$\neq 0$ may not be zero in general.

Therefore $u + v \notin U$. Hence set U is not a subspace of V_3.

In particular, if $(x_1, x_2, x_3) = u = (1, 0, 3)$, $v = (y_1, y_2, y_3) = (3, 2, -2) \in U$ since $0(1 + 3) = 0$, and $3(2 - 2) = 0$, but $y_2(x_1 + x_3) + x_2(y_1 + y_3) = 3(1 + 3) + 0(2 - 2) = 12 \neq 0$.

Hence U is not a subspace of V.

Example 14: Let $V = P_4$ be a vector space of polynomials of degree ≤ 4 with usual addition and scalar multiplication Let $U = \{p(x) \in P_4 \mid p''(1) = 2p'(1)\}$

To prove that U is a subspace of P_4, let $p(x), q(x) \in U$, then by definition, $(p+q)''(x) = p''(x) + q''(x)$ and $(p+q)''(1) = p''(1) + q''(1)$, since $(p+q)(x) = p(x) + q(x)$ $p''(1) = 2p'(1)$ and $q''(1) = 2q'(1)$.

To check the condition, we have $(p+q)''(1) = p''(1) + q''(1) = 2p'(1) + 2q'(1)$ $= 2(p'(1) + q'(1)) = 2(p + q)'(1) \Rightarrow (p + q) \in U$.

Also $\alpha p''(1) = \alpha 2 p''(1) = 2(\alpha p'(1))$. Therefore $(\alpha p)(x) \in U$.

Hence U is a subspace of P_4.

Example 15: Let $C'[a, b]$ be a set of continuously differentiable (derivative is continuous) real valued functions on the interval $[a, b]$. If a function is differentiable, then it is continuous also, but converse may not be true i.e. a continuous function may not be differentiable. Therefore $C'[a, b] \subset C[a, b]$ and $C'[a, b]$ is a subspace of $C[a, b]$ with the same rule of point-wise addition and point-wise scalar multiplication.

Similarly $C^{(n)}[a, b]$, n-times continuously differentiable functions on $[a, b]$ is a subspace of $C[a, b]$. Sum of two continuously differentiable function is continuously differentiable on the given interval and also scalar multiplication of a continuously differentiable function remains continuously differentiable function.

Therefore $C^{(n)}[a, b]$ is a subspace of $C[a, b]$ with the same rule of point-wise addition and point-wise scalar multiplication.

Example 16: Let $S = \{p(x) \in P_n(x) | \ p(1) + p(3) = 0\}$, $P_n(x)$ space of polynomials of degree $\leq n$. S is a subspace of $P_n(x)$.

To show that S is a subspace of $P_n(x)$, we have to check conditions (a) and (b) of Theorem 3.1.

First show that S is non empty, 0 polynomial $\in S$, since $0(1) + 0(3) = 0$.

(a) If $p(x), q(x) \in S$, then $p(1) + p(3) = 0$, $q(1) + q(3) = 0$

i.e. $(p + q)(1) + (p + q)(3) = p(1) + q(1) + p(3) + q(3)$

$= p(1) + p(3) + q(1) + q(3) = 0$, therefore $(p + q)(x) \in S$.

(b) If $p(x) \in S$, $p(1) + p(3) = 0 \Rightarrow \alpha(p(1) + p(3)) = \alpha 0 = 0$ for any scalar α. Therefore $\alpha p(x) \in S \Rightarrow (\alpha p)(x) \in S$.

Hence S is a subspace of $P_n(x)$, the space of polynomials of degree $\leq n$.

Example 17: Let $S = \{p(x) \in P_n(x) \ p(1) + p'(1) = 0\}$, S is a subspace of $P_n(x)$. Where $P_n(x)$ space of polynomials of degree $\leq n$.

Let $p(x), q(x) \in S$. as given $p(1) + p'(1) = 0$ and $q(1) + q'(1) = 0$

$(p + q)(x) = p(x) + q(x) \Rightarrow (p + q)(1) + (p + q)'(1) = p(1) + q(1) + p'(1) + q'(1)$

$= p(1) + p'(1) + q(1) + q'(1) = 0 + 0 = 0. \Rightarrow p + q \in S$.

$(\alpha p)(x) = \alpha p(x) \Rightarrow (\alpha p)(1) + (\alpha p)'(1) = \alpha(p(1) + p'(1)) = \alpha 0 = 0$

$\Rightarrow \alpha p \in S$. Hence S is a subspace of $P_n(x)$.

Theorem 3.2: Let U and W be any two subspaces of a vector space V, then $U \cap W$ is also a subspace of V.

Proof: Clearly $U \cap W$ is a non-empty subset of V, since U and W are subspaces of V, both U and W have 0 vector.

Let $u, v \in U \cap W$, then $u, v \in U$ and $u, v \in W$, by definition of intersection.

Since U and W are subspaces of V, $u + v \in U$, $\alpha u \in U$ and $u + v \in W$, $\alpha u \in W$.

Therefore $u + v \in U \cap W$, and $\alpha u \in U \cap W$ by definition of intersection.

Hence $U \cap W$ is a subspace of V.

Remark: Any intersection of subspaces of a vector space V is a subspace of V. However union of any two subspaces U and W of V need not be a subspace of vector space V i.e. $U \cup W$ need not be a subspace of V in general. This fact is illustrated by the following example.

Example 18: Let $U = \{(x, 0, 0), x - \text{real number}\}$ and $W = \{(0, y, 0), y - \text{real number}\}$ be two subspaces of vector space $V_3 = \{(x, y, z) - x, y, z \text{ any real numbers}\}$.

In this $U \cup W$ is set of vectors, which does not contain sum of a non-zero vector of U and a non-zero vector of W, because $\alpha(x, 0, 0) + \beta(0, y, 0) = (\alpha x, \beta y, 0) \notin U \cup W$ for all scalars α, β. Hence $U \cup W$ not a subspace of V in general.

Definition 3.4: Let U and W be two subspaces of a vector space V.

If $U \cap W = \phi$, then $U + W$ is denoted by $U \oplus W$ and read as direct sum of U and W.

3.3 LINEAR DEPENDENCE AND INDEPENDENCE

Definition 2.5 Let $S = \{u_1, u_2, ..., u_n\}$ be a subset of a vector space V. A vector $\alpha_1 u_1 + \alpha_2 u_2 +, ..., + \alpha_n u_n$ for some scalars $\alpha_1, \alpha_2, ..., \alpha_n$ is called a linear combination of a set $S = \{u_1, u_2, ..., u_n\}$.

(a) If all scalars $\alpha_1, \alpha_2, ..., \alpha_n$ are zeros, then linear combination is called **trivial**.

In other words if $|\alpha_1| + |\alpha_2| + ... |\alpha_n| = 0$, then linear combination is called **trivial**.

(b) If at least one of α_i's is not zero, then it is called **non-trivial** linear combination.

In other words **if** $|\alpha_1| + |\alpha_2| + ... |\alpha_n| \neq 0$, then it is called **non-trivial** linear combination

Now consider a linear combination of vectors of a set $S = \{u_1, ..., u_n\}$ equated to zero vector, i.e., $\alpha_1 u_1 + \alpha_2 u_2 +, ..., + \alpha_n u_n = 0_V$, and solve for $\alpha_1, \alpha_2, ..., \alpha_n$.

(c) If only solution is $\alpha_1 = \alpha_2 = , ..., = \alpha_n = 0$, then set $S = \{u_1, u_2, ..., u_n\}$ is called linearly independent set in short **LI**.

(d) If there exists a non-trivial solution for $\alpha_1, \alpha_2, ..., \alpha_n$ i.e. at least one of α_i's is not zero, then set $S = \{u_1, u_2, ..., u_n\}$ is called linearly dependent in short **LD**.

Fact 1: Any set $S = \{0_V, u_1, u_2, ..., u_n\}$ containing zero vector 0_V is always LD. Consider $\alpha_0 \neq 0$, $\alpha_0 0_V + 0\alpha_1 + , ..., + 0\alpha_n = 0_V$, this linear combination is not trivial and in this all scalars $\alpha_0, 0, 0, ..., 0$ are also not zeros, since $\alpha_0 \neq 0$. Therefore S is LD.

Therefore any set containing zero vector is always LD.

Fact 2: A single non-zero vector u of a vector space V, is always linearly independent since, only solutions of $\alpha u = 0_v$ is $\alpha = 0$.

Remark: Empty set is linearly independent by convention.

Example 19: Let $S = \{2 + x^2, x - 1, x + x^2\}$ be a subset of P_2.

Check whether S is linearly dependent.

Consider linear combination $\alpha(2 + x^2) + \beta(x - 1) + \gamma(x + x^2) = 0$.

On rearranging in power set x, we get $2\gamma - \beta + (\beta + \gamma)x + (\alpha + \gamma)x^2 = 0$

$\Rightarrow 2\alpha - \beta = 0, \beta + \gamma = 0, \alpha + \gamma = 0$. From last two relation

$\alpha - \beta = 0$ and $2\alpha - \beta = 0 \Rightarrow \alpha = 0, \beta = 0, \gamma = 0$. Set S is LI

Example 20: Determine whether the Set $S = \{1 + x, x + x^2, x^2 + 1\}$ of P_2 i.e., vector space of polynomials of degree ≤ 2 is LI or LD.

Consider the linear combination of the vectors of S, $\alpha(1 + x) + \beta(x + x^2) + \gamma(x^2 + 1) = 0$.

On rearranging in powers of x, we get

$(\alpha + \gamma) + (\alpha + \beta)x + (\beta + \gamma)x^2 = 0$ for all $x \in R$

(**Note:** this is not a quadratic equation to be solved for x.)

Equating the coefficients of powers of x, we get $\alpha + \gamma = 0$, $\alpha + \beta = 0$, $\beta + \gamma = 0$.

On subtracting second from first $\beta - \gamma = 0$, and $\beta + \gamma = 0$ we get $\beta = \gamma = 0$ and then $\alpha = 0$.

All three scalars $\alpha, \beta, \gamma = 0$. Therefore Set S is linearly independent i.e. LI.

Example 21: Check whether the set $S = \{(1, 1, 0), (1, 0, 1), (0, 1, 1)\}$ is LI in V_3.

Consider the linear combination $\alpha(1, 1, 0) + \beta(1, 0, 1) + \gamma(0, 1, 1) = (0, 0, 0)$

$(\alpha, \alpha, 0) + (\beta, 0, \beta) + (0, \gamma, \gamma) = (0, 0, 0)$, by scalar multiplication.

$(\alpha + \beta, \alpha + \gamma, \beta + \gamma) = (0, 0, 0)$, by vector addition.

$\alpha + \beta = 0, \alpha + \gamma = 0$ and $\beta + \gamma = 0$, by equality of vectors.

On solving, $\alpha = \beta = \gamma = 0$.

Therefore set S is linearly independent.

Example 22: Let $S = \{(1, -1, 2), (2, 3, 1), (4, 5, 6)\}$ be a subset of V_3. To check linear independence of S.

Consider linear combination

$\alpha(1, -1, 2) + \beta(2, 3, 1) + \gamma(4, 5, 6) = (0, 0, 0)$

$(\alpha, -\alpha, 2\alpha) + (2\beta, 3\beta, \beta) + (4\gamma, 5\gamma, 6\gamma) = (0, 0, 0)$

On equating, $\gamma + 2\beta + 4\gamma = 0$, $-\alpha + 3\beta + 5\gamma = 0$ and $2\gamma + \beta + 6\gamma = 0$

On solving the above equations, we get $\alpha = 0$, $\beta = 0$, $\gamma = 0$.

Hence S in linearly independent.

Example 23: Find k such that $\{(2, -1, 3), (3, 4, -1), (k, 2, 1)\}$ is L.I.

Consider a linear combination of first given two vectors.

$\alpha(2, -1, 3) + \beta(3, 4, -1) = (2\alpha, -\alpha, 3\alpha) + (3\beta, 4\beta, -\beta) = (2\alpha + 3\beta, -\alpha + 4\beta, 3\alpha - \beta)$

On equating to third vector $(k, 2, 1)$, $2\alpha + 3\beta = k$, $-\alpha + 4\beta = 2$ and $3\alpha - \beta = 1$

On solving last two of the equations, we get $11\beta = 7$, $\beta = \dfrac{7}{11}$, $3\alpha = 1 + \dfrac{7}{11} = \dfrac{18}{11}$,

$\alpha = \dfrac{6}{11}$, so $k = 2\alpha + 3\beta = 2 \times \dfrac{6}{11} + 3 \times \dfrac{7}{11} = 3$.

If k is chosen 3, then set becomes LD.

Therefore for set to be LI, we should choose k any number other than 3, that is $k \ne 3$.

Example 24: Let $S = \{(1, -2, 3, -1), (2, 1, -1, 2), (3, -1, 2, 1)\}$ be a subset of R^4. To check whether set S is linearly independent.

Consider the linear combination $\alpha(1, -2, 3, -1) + \beta(2, 1, -1, 2) + \gamma(3, -1, 2, 1) = (0, 0, 0, 0)$.

$(\alpha, -2\alpha, 3\alpha, -\alpha) + \beta(2\beta, \beta, -\beta, 2\beta) + (3\gamma, -\gamma, 2\gamma, \gamma) = (0, 0, 0, 0)$ by scalar multiplication $(\alpha + 2\beta + 3\gamma, -2\alpha + \beta - \gamma, 3\alpha - \beta + 2\gamma, -\alpha + 2\beta + \gamma) = (0, 0, 0, 0)$ by addition of vectors.

$\alpha + 2\beta + 3\gamma = 0$, $-2\alpha + \beta - \gamma = 0$, $3\alpha - \beta + 2\gamma = 0$, $-\alpha + 2\beta + \gamma = 0$

$\alpha + 2\beta + 3\gamma = 0$ \hfill (1)

$-2\alpha + \beta - \gamma = 0$ \hfill (2)

$3\alpha - \beta + 2\gamma = 0$ \hfill (3)

$-\alpha + 2\beta + \gamma = 0$ \hfill (4)

From (1) and (2) $5\beta + 5\gamma = 0$, from (1) and (4) $4\beta + 4\gamma = 0$, from both $\beta + \gamma = 0$, using (3) $\alpha = \beta$, $\gamma = 1$, $\beta = -1$, $\alpha = -1$.

Hence set is linearly dependent.

Example 25: Let $\{u, v, w, z\}$ be a set of linearly independent vectors of a vector space. Check whether the set $S = \{u - 3z, v + 2u, 2v - w, w + z\}$ is linearly independent.

Consider the linear combination $\alpha(u - 3z) + \beta(v + 2u) + \gamma(2v - w) + \delta(w + z) = 0_V$.

On rearranging $(\alpha + 2\beta)u + (\beta + 2\gamma)v + (\delta - \gamma)w + (\delta - 3\alpha)z = 0_v$.

Since $\{u, v, w, z\}$ is L.I.

$$\alpha + 2\beta = 0, \ \beta + 2\gamma = 0, \ \delta - \gamma = 0, \text{ and } \delta - 3\alpha = 0.$$

we get $\alpha + 2\beta = 0, \ \beta + 2\gamma = 0, \ \delta - \gamma = 0, \ \delta - 3\alpha = 0. \Rightarrow \delta = \gamma, \ \delta = 3\alpha$

Further on solving, we get $\alpha = \beta = \gamma = \delta = 0$.

Hence set $S = \{u - 3z, v + 2u, 2v - w, w + z\}$ is linearly independent.

Example 26: Check whether set $S = \{(3, -2, 5), (1, 1, 0), (1, 0, 1), (0, 1, 1)\}$ in V_3 is LI.

Consider the linear combination

$\alpha(3, -2, 25) + \beta(1, 1, 0) + \gamma(1, 0, 1) + \delta(0, 1, 1) = (0, 0, 0)$

$(3\alpha, -2\alpha, 5\alpha) + (\beta, \beta, 0) + (\gamma, 0, \gamma) + (0, \delta, \delta) = (0, 0, 0)$, by scalar multiplication.

$(3\alpha + \beta + \gamma, -2\alpha + \beta + \delta, 5\alpha + \gamma + \delta) = (0, 0, 0)$, by addition of vectors.

$3\alpha + \beta + \gamma = 0, -2\alpha + \beta + \delta = 0$, and $5\alpha + \gamma + \delta = 0$. By equality of vectors

On solving these equations, $\beta = 2\alpha, \ \delta = 0, \ \gamma = -5\alpha$, we note that for different values of α, we have many non-trivial solutions.

Therefore set S is Linearly Dependent.

Example 27: Given that set $\{u, v, w\}$ is Linearly Independent, in a vector space V, check whether the set $\{u - v, v - w, w - u\}$ is LI.

Consider $\quad \alpha(u - v) + \beta(v - w) + \gamma(w - u) = 0_v$

on rearranging $\quad (\alpha - \gamma)u + (\beta - \alpha)v + (\gamma - \beta)w = 0_v$

Since $\quad \{u, v, w\}$ is L.I. $\quad \alpha - \gamma = \beta - \alpha = \gamma - \beta = 0$

$\Rightarrow \alpha = \beta = \gamma$, which may be other than zeros. Therefore many non-trivial linear combinations are obtained. Given set is L.D.

Theorem 3.3:

(a) If set $S = \{u_1, u_2, ..., u_k\}$ of a vector space V is L.D, then every superset of S is also L.D.

(b) If a set $S = \{u_1, u_2, ..., u_n\}$ of a vector space V is linearly independent, then every subset of S is also linearly independent.

Proof:

(a) Let $S = \{u_1, u_2, ..., u_k\}$ be a linearly dependent set of a vector space V.

Consider linear combination $\alpha_1 u_1 + \alpha_2 u_2 + ... + \alpha_k u_k = 0_v$

Set S is linearly dependent, therefore there is at least one of α's which is non-zero say $\alpha_i \neq 0, \ i \geq k$

Now consider a superset $S_1 = \{u_1, u_2, ..., u_k, u_{k+1}, ..., u_n\}$ of S.
Extend above linear combination to
$$a_1 u_1 + a_2 u_2 ... + a_k u_k + 0 u_{k+1} + ... + ... 0 u_n = 0_v.$$
These α's are same, and one of α's is not zero, therefore set
$S_1 = \{u_1, u_2, ..., u_k, u_n, ..., u_n\}$ is also LD

(b) If a set $S = \{u_1, u_2, ..., u_n\}$ of a vector space V is linearly independent.

Consider $S_1 = \{u_1, u_2, ..., u_k\}$ a subset of any k vectors of S, $k \le n$ (these k vectors may not be necessarily first k vectors, but suffices may be rearranged).

Suppose S_1 is not L.I., i.e., L.D. From (a) every superset of S_1 is L.D., therefore S is L.D., but this contradicts that S is L.I. Hence assumption that S_1 is not L.I., i.e., L.D. is false. Therefore S_1 is linearly independent.

Definition 3.6: **Span of a set:** Consider a nonzero element u_0 of a vector space V. Now all vectors au_0 for all real numbers α make a subspace of V, namely, $[u_0]$, which is fully identified, once we know the single element u_0. In this section subspaces that are fully identified by a subset of V, which is smaller than the subspace are to be studied.

Let $S = \{u_1, u_2, ..., u_n\}$ be a subset of a vector space V. Set of all vectors, which are linear combinations of S, i.e. $u = \alpha_1 u_1 + \alpha_2 u_2 + ... + \alpha_n u_n$, where $\alpha_1, \alpha_2,..., \alpha_n$ vary over all real numbers, is a set of infinite vectors, since α_1 will take up any real value, α_2 will take any real value, and so on. Such a set is called span of S, and denoted by
$$[S] = [u_1, u_2, ...u_n] = \{u = a_1 u_1 + ... + a_r u_n, \text{ for all real numbers } \alpha_1, \alpha_2, ... \alpha_n\}.$$

Example 28: Check whether $(3, 2, 4) \in [(1, 2, 1), (-1, 3, 2), (2, -1, 1)]$.

Consider the linear combination $(3, 2, 4) = \alpha(1, 2, 1) + \beta(-1, 3, 2) + \gamma(2, -1, 1)$.

If we can find the values of the numbers α, β, γ satisfying the above, the given vector is in the span of the set.

On equating, we get $\alpha - \beta + 2\gamma = 3$, $2\alpha + 3\beta - \gamma = 2$, and $\alpha + 2\beta + \gamma = 4$.

On solving these equations, we get $\alpha = 5/10$, $\beta = 9/10$, $\gamma = 17/10$.

Hence $(3, 2, 4) \in [(1, 2, 1), (-1, 3, 2), (2, -1, 1)]$ is correct.

Example 29: Check whether $(3, 2, 5) \in [(1, 2, 1), (-1, 3, 2), (2 -1, -1)]$.

Consider the linear combination $(3, 2, 5) = \alpha(1, 2, 1) + \beta(-1, 3, 2) + \gamma(2, -1, -1)$

If we can find the values of the numbers α, β, γ satisfying the above, the vector is in the span of the given set.

On equating, we get $\alpha - \beta + 2\gamma = 3$, $2\alpha + 3\beta - \gamma = 2$, and $\alpha + 2\beta - \gamma = 5$.

On solving these equations, we get $\alpha + \beta = -3$ and $\alpha + \beta = 3$, which is inconsistent. Hence no solution exist and. $(3, 2, 5) \notin [(1, 2, 1), (-1, 3, 2), (2, -1, -1)]$.

If we solve the above equations by row reduction method, then

Augmented matrix is $\begin{bmatrix} 1 & -1 & 2 & 3 \\ 2 & 3 & -1 & 2 \\ 1 & 2 & -1 & 5 \end{bmatrix}$

$r_2 \leftarrow r_2 - 2r_1, r_3 \leftarrow r_3 - r_1 \sim \begin{bmatrix} 1 & -1 & 2 & 3 \\ 0 & 5 & -5 & -4 \\ 0 & 3 & -3 & 2 \end{bmatrix}$

$r_2 \leftarrow \frac{1}{5} r_2, r_3 \leftarrow \frac{1}{3} r_3 \sim \begin{bmatrix} 1 & -1 & 2 & 3 \\ 0 & 1 & -1 & -0.8 \\ 0 & 0 & 0 & 1.4667 \end{bmatrix}$

Rank $A = 2$, rank $(A, b) = 3$, therefore rank $A \neq$ rank (A, b). Therefore system is inconsistent, No solution exists. Hence $(3, 2, 5)$ does not belong to the span.

Example 30: Check whether $x^2 + 3x + 7 \in [1 - x, x - x^2, 1 + x^2]$.

Consider the linear combination

$$x^2 + 3x + 7 = \alpha(1 - x) + \beta(x - x^2) + \gamma(1 + x^2)$$
$$= (\alpha + \gamma) + (-\alpha + \beta)x + (\gamma - \beta)x^2$$

on equating the coefficients of power of x,

we get $(\alpha + \gamma) = 7$, $(-\alpha + \beta) = 3$ and $(\gamma - \beta) = 1$.

On solving the equations, we get $\alpha = \frac{3}{2}$, $\beta = \frac{3}{2}$, $\gamma = \frac{3}{2}$.

Hence $x^2 + 3x + 7 \in [1 - x, x - x - x^2, 1 + x^2]$.

Remark: If a set B consists of infinite number of vectors of a vector space, then B is called linearly independent if every finite subset of B is linearly independent.

Theorem 3.4: If S is a non-empty finite subset of a vector space V, then $[S]$ is a subspace of V and $[S]$ is the smallest subspace of vector space V, containing S.

Let $S = \{u_1, u_2, ..., u_n\}$ be a subset of V. $[S]$ is shown closed for addition and scalar multiplication for $[S]$ to be proved a subspace of V

Let $u = \alpha_1 u_1 + \alpha_2 u_2 + ... + \alpha_n u_n$, and $v = b_1 u_1 + b_2 u_2 + ... + b_n u_n$ for some scalars α_i's, β_i's be any two vectors in $[S]$.

$$u + v = \alpha_1 u_1 + \alpha_2 u_2 + ... + \alpha_n u_n + \beta_1 u_1 + \beta_2 u_2 + ... + \beta_n u_n$$
$$= (\alpha_1 + \beta_1)u_1 + (\alpha_2 + \beta_2)u_2 + ... + (\alpha_n + \beta_n)u_n$$

is a linear combination of $\{u_1, u_2, ..., u_n\}$.

Vector Spaces

Therefore $u + v \in [S]$.

Similarly, $\alpha u = \alpha(\alpha_1 u_1 + \alpha_2 u_2 + ... + \alpha_n u_n) = \alpha\alpha_1 u_1 + \alpha\alpha_2 u_2 + ... + \alpha\alpha_n u_n$ is again a linear combination of vectors of S and so it is in $[S]$. Hence, $[S]$ is a subspace of V.

Clearly $[S]$ contains S, because each element u_1 of S can be written as $1u_1$, i.e. a finite linear combination of S. To prove that $[S]$ is the smallest subspace containing S, we shall show that if there exists another subspace T containing S, then T contains $[S]$ also.

So let a subspace T contain S. We have to prove that T contains $[S]$. Take any element of $[S]$. It is of the form $\alpha_1 u_1 + \alpha_2 u_2 + ... + \alpha_n u_n$, where αi 's are scalars.. Since $S \subset T$, each u_i also belongs to T. Since T is a subspace, $\alpha_1 u_1 + \alpha_2 u_2 + ... + \alpha_n u_n$ should also belong to T. This means that each element of $[S]$ is in T. This proves T contains $[S]$.

Hence $[S]$ is the smallest subspace containing S.

Remark: A nontrivial subspace always contains an infinite number of elements. So $[S]$ ($\neq V_0$) always contains an infinite number of elements. But S itself may be a smaller set. By convention $[\phi] = V_0$ is taken.

Example 31: Let $\{v_1, v_2\}$ be a subset of a vector space V. To prove that

$$[v_1, v_2] = [v_1 - v_2, v_1 + v_2],$$

Let $u \in [v_1, v_2]$, $u = \alpha v_1 + \beta v_2$ for some scalars α, β.

Now u can be written as $u = \dfrac{\alpha - \beta}{2}(v_1 - v_2) + \dfrac{\alpha + \beta}{2}(v_1 + v_2)$, this show that $u \in [v_1 - v_2, v_1 + v_2]$. Hence $[v_1, v_2] \subset [v_1 - v_2, v_1 + v_2]$.

Now Let $v \in [v_1 - v_2, v_1 + v_2]$

$v = \alpha(v_1 - v_2) + \beta(v_1 + v_2) = (\alpha + \beta)v_1 + (\beta - \alpha)v_2$ for some scalars α and β.

$u \in [v_1, v_2]$ is arbitrary. Hence $[v_1 - v_2, v_1 + v_2] \subset [v_1, v_2]$.

Therefore $[v_1, v_2] = [v_1 - v_2, v_1 + v_2]$.

Example 32: Let $S \subset V$, S be a subset of vector space V and $u, v \in V$, if $u \in [S \cup \{v\}]$ but $u \notin [S]$, then prove that $v \in [S \cup \{u\}]$.

Given $u \in [S \cup \{v\}]$, $u \in [S] + \beta v$, but $u \notin [S]$, then $\beta \neq 0$ if $\beta = 0$ then $u \in [S]$, which contradicts the given condition $u \notin [S]$. Therefore $u = \alpha_1 u_1 + , ..., + \alpha_n u_n + \beta v$ for some $u_1, u_2, ..., u_n \in S$ and scalars $\alpha_1, \alpha_2, ..., \alpha_n$, with $\beta \neq 0$.

So $\beta v = u - \alpha_1 u_1, ..., \alpha_n u_n$, since $\beta \neq 0$, $v = \dfrac{1}{\beta}u - \dfrac{\alpha_1}{\beta}u_1, ..., -\dfrac{\alpha_n}{\beta}u_n$.

This shows that $v \in [S \cup \{u\}]$. Hence proved.

Theorem 2.5: Suppose $S = \{v_1, v_2, ..., v_k\}$ is an ordered set of a vector space V.

If $v_1 \neq 0$, then set S is LD, iff one of the vectors of $\{v_1, ..., v_k\}$ belongs to the span of remaining other vectors of set S.

Let $v_k \in [v_1, v_2, ..., v_{k-1}]$. There are scalars $\alpha_1, \alpha_2, ..., \alpha_{k-1}$ such that
$$v_k = \alpha_1 v_1, \alpha_2 v_2, ..., \alpha_{k-1} v_{k-1}$$
i.e.,
$$\alpha_1 v_1 + \alpha_2 v_2 + ... + \alpha_{k-1} v_{k-1} - v_k = 0.$$

In this linear combination, scalar -1 with v_k is non-zero.

This shows that a non-trivial linear combination of $\{v_1, v_2, ..., v_{k-1}, v_k\}$ is 0_V.

Hence set S is L.D.

Conversely, if $S = \{v_1, v_2, ..., v_k\}$ is L.D., then for some non-zero scalars $\alpha_1, \alpha_2, \alpha_3, ..., \alpha_n$
$$\alpha_1 v_1 + \alpha_2 v_2 + ... + \alpha_k v_k = 0_v \quad \text{(say } \alpha_k \neq 0\text{)}.$$

On writing
$$\alpha_k v_k = -\alpha_1 v_1 - \alpha_2 v_2 - ... - \alpha_{k-1} v_{k-1}.$$

Since $\alpha_k \neq 0$,
$$v_k = -\frac{\alpha_1}{\alpha_k} v_1 - \frac{\alpha_2}{\alpha_k} v_2, ..., -\frac{\alpha_{n-1}}{\alpha_k} v_{k-1}$$

i.e., v_k is linear combination of vectors $\{v_1, v_2, ..., v_{k-1}\}$.

Therefore $v_k \in \{v_1, v_2, ..., v_{k-1}\}$. Hence proved.

Theorem 2.6: Let $B = \{u_1, u_2, ..., u_k\}$ spans V. The statements
(a) Set B is linearly independent,
(b) $u \in V$ is expressed uniquely as $u = \alpha_1 u_1 + \alpha_2 u_2 + ... + \alpha_n u_n$, are equivalent

Assume (a). We shall prove that any expression for u in terms of $u_1, u_2, ..., u_n$ is unique.

For, if
$$u = \alpha_1 u_1 + \alpha_2 u_2 + ... + a_n u_n \qquad (1)$$
And also
$$u = \beta_1 u_1 + \beta_2 u_2 + ... + \beta_n u_n \qquad (2)$$
Then
$$\alpha_1 u_1 + \alpha_2 u_2 + ... + \alpha_n u_n = \beta_1 u_1 + \beta_2 u_2 + ... + \beta_n u_n.$$
$$(\alpha_1 - \beta_1) u_1 + (\alpha_2 - \beta_2) u_2 + ... + (\alpha_n - \beta_n) u_n = 0_V$$

Since u_i's are L.I, therefore $\alpha_i - \beta_i = 0$ for all $i = 1, 2, ... n$.

This gives $\alpha_i = \beta_i$ for all i. Hence expression (1) is unique.

Conversely assume (b). Suppose now that the u_i's are not L.I, i.e. they are LD. Then there exists a nontrivial linear combination, say
$$\alpha_1 u_1 + \alpha_2 u_2 + ... + \alpha_n u_n = 0_V, \text{ which equals the zero vector,}$$
also $0 u_1 + 0 u_2 + ... + 0 u_n = 0_v.$

Thus, we get two different expressions for 0_v. This contradicts (b).

Hence $\{u_1, u_2, ..., u_n\}$ is L.I.

Many times while checking linear dependence or independence of a given set, solution of the equations involving scalars may not be simple. In such cases, use of row reduced echelon form may be more convenient. Therefore some problems of checking linear independence are solved by row reduced echelon method.

Linear Dependence and Independence can also be Checked by Row Reduced Echelon Form

Row Reduction method: Let $S = \{u_1, u_2, ..., u_m\}$ be a subset of n-truple vector space Vn. If $m > n$, then set S is L.D., because number of vectors is more than the dimension of $R^n = V_n$. Therefore for $m \leq n$, linear independence of S can be checked by using row-reduced echelon-form.

Suppose $u_i = (x_{i1}, x_{i2}, ..., x_{in})$, $i = 1, 2 ... m$ be m vectors in V_n.

Now consider $m \times n$ matrix A.

$$\begin{bmatrix} x_{11} & x_{12} & - & - & - & - & x_{1n} \\ x_{21} & x_{22} & - & - & - & - & x_{2n} \\ - & - & - & - & - & - & - \\ - & - & - & - & - & - & - \\ - & - & - & - & - & - & - \\ x_{m1} & x_{m2} & - & - & - & - & x_{mn} \end{bmatrix}$$

Components of the vectors have been written as rows in the above matrix. Above matrix is reduced to row reduced echelon form. All the rows, which are reduced to zero rows, have been linear combination of non-zero rows, therefore if rank $A = r$, the number of non-zero rows, then the number of linearly independent vectors would be r.

Also $r \leq m \leq n$. Therefore, steps are

(1) write the vectors row-wise,

(2) reduce to row reduced echelon form;

(3) number of non-zero rows is the number of linearly independent vectors.

The transpose of the above matrix will also give the same result i.e. vectors are written column-wise. If S is linearly independent set, then solution of $\alpha_1 u_1 + \alpha_2 u_2 + ... \alpha_m u_m = 0$ should be trivial.

Augmented matrix for $\alpha_1, \alpha_2 \ldots \alpha_m$ is

$$\begin{bmatrix} x_{11} & x_{21} & - & - & - & x_{m1} & 0 \\ x_{12} & x_{22} & - & - & - & x_{m2} & 0 \\ - & - & - & - & - & - & - \\ - & - & - & - & - & - & - \\ - & - & - & - & - & - & - \\ x_{1n} & x_{2n} & - & - & - & x_{mn} & 0 \end{bmatrix}, m \leq n.$$

On reducing the above matrix to row reduced echelon form

$$\begin{bmatrix} 1 & 0 & - & - & 0 & 0 \\ 0 & 1 & 0 & - & 0 & 0 \\ 0 & 0 & 1 & - & 0 & 0 \\ - & - & - & - & - & - \\ 0 & 0 & 0 & - & 1 & 0 \end{bmatrix}.$$

If rank = m i.e. first m columns will form unit matrix, and last column of zeros can ignored.

If rank < m, then $S = \{u_1, u_2, \ldots u_m\}$ is LD, if rank = m, then S is L.I.

Example 33: To find the number of linearly independent vectors of the given set

$$S = \{(2, -1, 3, 1)\}, (1, 2, 1, 3), (3, 1, 4, 4), (-1, 3, -2, 2)\},$$

Above vectors written as rows of the matrix

$$A = \begin{bmatrix} 2 & -1 & 3 & 1 \\ 1 & 2 & 1 & 3 \\ 3 & 1 & 4 & 4 \\ -1 & 3 & -2 & 2 \end{bmatrix},$$ and this would be reduced to row reduced echelon form.

Rank of the above matrix will give the number of linearly independent vectors.

On $r_1 \leftarrow r_1 - 2r_2$, $r_3 \leftarrow r_3 - 3r_2$, and $r_4 \leftarrow r_4 + r_2$,

$$A \sim \begin{bmatrix} 0 & -5 & 1 & -5 \\ 1 & 2 & 1 & 3 \\ 0 & -5 & 1 & -5 \\ 0 & 5 & -1 & 5 \end{bmatrix}. \text{ On } r_1 \leftrightarrow r_2 \sim \begin{bmatrix} 1 & 2 & 1 & 3 \\ 0 & -5 & 1 & -5 \\ 0 & -5 & 1 & -5 \\ 0 & 5 & -1 & 5 \end{bmatrix}$$

$r_4 \leftarrow r_4 + r_2, \; r_3 \leftarrow r_3 - r_2 \sim \begin{bmatrix} 1 & 2 & 1 & 3 \\ 0 & -5 & 1 & -5 \\ 0 & 0 & 0 & 0 \\ 0 & 0 & 0 & 0 \end{bmatrix}$

$r_2 \leftarrow -\dfrac{1}{5} \times r_2 \sim \begin{bmatrix} 1 & 2 & 1 & 3 \\ 0 & 1 & -\dfrac{1}{5} & 1 \\ 0 & 0 & 0 & 0 \\ 0 & 0 & 0 & 0 \end{bmatrix}$, and $r_1 \leftarrow r_1 - 2r_2 \sim \begin{bmatrix} 1 & 0 & \dfrac{7}{5} & 1 \\ 0 & 1 & -\dfrac{1}{5} & 1 \\ 0 & 0 & 0 & 0 \\ 0 & 0 & 0 & 0 \end{bmatrix}$

Rank $A = 2$, Hence given vectors are L.D., but two vectors are L.I., Therefore dim $[S] = 2$. (**Note:** dim is defined in Section 2.4)

Example 34: Now writing the vectors of above matrix column wise:

$B \sim \begin{bmatrix} 2 & 1 & 3 & -1 \\ -1 & 2 & 1 & 3 \\ 3 & 1 & 4 & -2 \\ 1 & 3 & 4 & 2 \end{bmatrix}$, on $r_1 \leftrightarrow r_4 \sim \begin{bmatrix} 1 & 3 & 4 & 2 \\ -1 & 2 & 1 & 3 \\ 3 & 1 & 4 & -2 \\ 2 & 1 & 3 & -1 \end{bmatrix}.$

On $r_2 \leftarrow r_2 + r_1, \; r_3 \leftarrow r_3 - 3r_1, \; r_4 \leftarrow r_4 - 2r_1.$

$\sim \begin{bmatrix} 1 & 3 & 4 & 2 \\ 0 & 5 & 5 & 5 \\ 0 & -8 & -8 & -8 \\ 0 & -5 & -5 & -5 \end{bmatrix} \sim \begin{bmatrix} 1 & 3 & 4 & 2 \\ 0 & 1 & 1 & 1 \\ 0 & 0 & 0 & 0 \\ 0 & 0 & 0 & 0 \end{bmatrix} \sim \begin{bmatrix} 1 & 0 & 1 & -1 \\ 0 & 1 & 1 & 1 \\ 0 & 0 & 0 & 0 \\ 0 & 0 & 0 & 0 \end{bmatrix}$

This is not unit matrix, only two non-zero rows are obtained. Hence number of linearly independent vectors is 2.

Example 35: Check whether $(3, 2, -4, 5) \in [(1, 2, 3, -1), (2, -3, 2, 5)]$.

The two vectors are linearly independent, which spans the space. Therefore if given vector belongs to the span, then set should be L.D.

3.24

Consider the matrix of rows as obtained from vectors:

$$\begin{bmatrix} 1 & 2 & 3 & -1 \\ 2 & -3 & 2 & 5 \\ 3 & 2 & -4 & 5 \end{bmatrix} \text{ on } r_2 \leftarrow r_2 - 2r_1,\ r_3 \leftarrow r_3 - 3r_1 \sim \begin{bmatrix} 1 & 2 & 3 & -1 \\ 0 & -5 & -4 & 7 \\ 0 & -4 & -13 & 8 \end{bmatrix}$$

$$r_2 \leftarrow -\frac{1}{5} \times r_2,\ r_3 \leftarrow -\frac{1}{4} \times r_3 \sim \begin{bmatrix} 1 & 2 & 3 & -1 \\ 0 & 1 & \frac{4}{5} & -\frac{7}{5} \\ 0 & 1 & \frac{13}{4} & -2 \end{bmatrix}$$

$$r_1 \leftarrow r_1 - 2r_2,\ r_3 \leftarrow r_3 - r_2 \sim \begin{bmatrix} 1 & 0 & \frac{7}{5} & \frac{9}{5} \\ 0 & 1 & \frac{4}{5} & -\frac{7}{5} \\ 0 & 0 & \frac{49}{20} & -\frac{3}{5} \end{bmatrix}$$

$$r_3 \leftarrow \frac{20}{49} r_3 \sim \begin{bmatrix} 1 & 0 & \frac{7}{5} & \frac{9}{5} \\ 0 & 1 & \frac{4}{5} & -\frac{7}{5} \\ 0 & 0 & 1 & -\frac{8}{48} \end{bmatrix}$$

$$r_2 \leftarrow r_2 - \frac{4}{5} r_3,\ r_1 \leftarrow r_1 - \frac{7}{5} r_3 \sim \begin{bmatrix} 1 & 0 & 0 & \frac{497}{245} \\ 0 & 1 & 0 & -\frac{311}{245} \\ 0 & 0 & 1 & -\frac{8}{49} \end{bmatrix}.$$

Non-zero rows are 3, therefore rank of the matrix is 3. All three vectors are linearly independent.

Hence given vector is not in the span of the other two vectors.

Vector Spaces

Example 36: Check whether the set $S = \{x - 1, x + 2x^2, x^3 + 1, 2x^3 - 3x\}$ of P_3 is linearly independent?

On arranging the coefficients of increasing powers of x of the vector polynomials of set S in matrix form as given below:

$$\begin{array}{cccc} 1 & x & x^2 & x^3 \end{array}$$
$$\begin{bmatrix} -1 & 1 & 0 & 0 \\ 0 & 1 & 2 & 0 \\ 1 & 0 & 0 & 1 \\ 0 & -3 & 0 & 2 \end{bmatrix}.$$

Now to reduce to row–reduced echelon form on $r_1 \leftrightarrow r_3 \sim \begin{bmatrix} 1 & 0 & 0 & 1 \\ 0 & 1 & 2 & 0 \\ -1 & 1 & 0 & 0 \\ 0 & -3 & 0 & 2 \end{bmatrix}.$

$r_3 \leftarrow r_3 + r_1 \sim \begin{bmatrix} 1 & 0 & 0 & 1 \\ 0 & 1 & 2 & 0 \\ 0 & 1 & 0 & 1 \\ 0 & -3 & 0 & 2 \end{bmatrix}, \quad r_2 \leftrightarrow r_3 \sim \begin{bmatrix} 1 & 0 & 0 & 1 \\ 0 & 1 & 0 & 1 \\ 0 & 1 & 2 & 1 \\ 0 & -3 & 0 & 2 \end{bmatrix}.$

$r_3 \leftarrow r_3 - r_2, r_4 \leftarrow r_4 + 3r_2 \sim \begin{bmatrix} 1 & 0 & 0 & 1 \\ 0 & 1 & 0 & 1 \\ 0 & 0 & 2 & 0 \\ 0 & 0 & 0 & 5 \end{bmatrix},$

$r_3 \leftarrow \dfrac{1}{2}r_3, r_4 \leftarrow \dfrac{1}{5}r_4 \sim \begin{bmatrix} 1 & 0 & 0 & 1 \\ 0 & 1 & 0 & 1 \\ 0 & 0 & 1 & 0 \\ 0 & 0 & 0 & 1 \end{bmatrix}.$

$r_1 \leftarrow r_1 - r_4, r_2 \leftarrow r_2 - r_4 \sim \begin{bmatrix} 1 & 0 & 0 & 0 \\ 0 & 1 & 0 & 0 \\ 0 & 0 & 1 & 0 \\ 0 & 0 & 0 & 1 \end{bmatrix}.$

Number of non-zero rows is 4, therefore number of linearly independent vectors is 4. Hence the set is linearly independent.

3.4 BASIS AND DIMENSION

The set of vectors $\{e_1, e_2, ..., e_n\}$, where $e_i = (0, 0, ..., 1, ..., 0, 0)$, 1 at i^{th} place, $i = 1, 2, ..., n$ is both linearly independent and spans R^n (or V_n). Such sets are very important in the theory of vector spaces and hence we have the following definition:

Definition 3.7: Let V be a vector space. A subset B of V is called a basis of V, if

(a) B is Linearly Independent set.

(b) $[B] = V$.

Remark: There may be many subsets of V, which are bases of V.

Definition 3.8: Let B be a basis of a vector space V. If the number of vectors in B is n, then vector space V is called n-dimensional and written as dim $V = n$.

Dimension of trivial vector space $V_0 = \{0_V\}$ is taken to be zero.

If n is finite, then vector space V is called finite dimensional and this number n is same for all bases of V. If n is infinite, then vector space V is called infinite dimensional.

Here after only finite dimensional vector spaces would be considered mostly.

The checking of linear dependence or linear independence of a subset S of a finite dimensional vector space in some cases can be determined easily, if the following criterion is used.

Let $B = \{u_1, u_2, ..., u_n\}$ be a basis of an n-dimensional vector space V. Take any vector $v \in V$, by definition of a basis, $v = \alpha_1 u_1 + \alpha_2 u_2 + ... + \alpha_n u_n$, for some scalars $\alpha_1, \alpha_2, ..., \alpha_n$. This shows that set $\{u_1, u_2, ..., u_n v\}$ is L.D.

Therefore in general any subset S of $(n + 1)$ vectors of an n-dimensional vector space is always L.D. Continuing similarly a set $S = \{u_1, u_2, ..., u_m\}$ of n-dimensional vector space V is L.D. if $m > n$ in general.

Corollary: If V has a basis of n elements, then every set of p vectors of V, with $p > n$, is L.D.

Fact 1: If V has a basis of n elements, then every other basis also has n elements.

Let $B = \{u_1, u_2, ..., u_n\}$ be a basis of an n-dimensional vector space V, and $B_1 = \{u_1, u_2, ..., u_m\}$ be another basis of V. If $m > n$, then B_1 is L.D. If $n > m$, then B is L.D. Hence $m = n$.

Fact 2: In an n-dimensional vector space V, any set of n linearly independent vectors of V is basis of V.

Example 37: Let $B = \{(1, 0, 0), (0, 1, 0), (0, 0, 1)\}$ be a subset of V_3.

Consider linear combination $\alpha(1, 0, 0) + \beta(0, 1, 0) + \gamma(0, 0, 1) = (0, 0, 0)$, with α, β, γ scalars.

$(\alpha, 0, 0) + (0, \beta, 0) + (0, 0, \gamma) = (0, 0, 0)$ by scalar multiplication,

$(\alpha, \beta, \gamma) = (0, 0, 0)$ by addition of vectors,

$\alpha = \beta = \gamma = 0$ by equality of vectors. Hence B is linearly independent.

Now take any vector $(x_1, x_2, x_3) \in V_3$.

Consider $(x_1, x_2, x_3) = \alpha(1, 0, 0) + \beta(0, 1, 0) + \gamma(0, 0, 1)$. Our aim is to check whether (x_1, x_2, x_3) is a linear combination of B.

$(x_1, x_2, x_3) = (\alpha, \beta, \gamma)$ by vector addition and scalar multiplication.

On equating $x_1 = \alpha$, $x_2 = \beta$, $x_3 = \gamma$. Hence B is a basis.

This shows that (x_1, x_2, x_3) is always a linear combination of B for any real numbers x_1, x_2, x_3.

Therefore $[B] = V_3$, B is a basis of V_3, and dim $V_3 = 3$. This B is called a standard basis of V_3

Example 38: Let $B = \{1, x, x^2, x^3\}$ be a subset of P_3, the space of polynomials of degree ≤ 3. To show that B is a basis of P_3.

To check B is L.I, consider $\alpha + \beta x + \gamma x^2 + \delta x^3 = 0$ for all $x \in R$.

(Note: This is not a cubic equation in valuable x for given values of $\alpha, \beta, \gamma, \delta$).

Our aim to find the values of $\alpha, \beta, \gamma, \delta$ such that above is true for all values of $x \in R$.

On right side 0 should be considered zero polynomial, then equating the coefficients of powers of x, we get $\alpha = \beta = \gamma = \delta = 0$. Hence B is L.I.

Now to check that B spans P_3, take any polynomial $p(x) = \alpha_0 + \alpha_1 x + \alpha_2 x^2 + \alpha_3 x^3$. Obviously $p(x)$ is linear combination of B for real numbers $\alpha_0, \alpha_1, \alpha_2$, and α_3.

Hence B is a basis of P_3 and dim $P_3 = 4$.

In general $\{1, x, x^2 \ldots x^n\}$ is called standard basis of P_n and dim $P_n = n + 1$.

Example 39: Let U be a subspace of vector space P_2 defined by

$$U = \{p(x) \in P_2 \mid p(1) = 0\}. \text{ Find a basis of } U.$$

Take a polynomial $p(x) = a_0 + a_1 x + a_2 x^2 \in P_2$.

For $p(x) \in U$, $p(1) = 0 = a_0 + a_1 + a_2 \Rightarrow a_0 = -(a_1 + a_2)$,

Now replacing a_0 by $-(a_1 + a_2)$, (any of three a_0, a_1, a_2 can be replaced in terms of other two, therefore answers may also appear in different forms)

$$p(x) = -(a_1 + a_2) + a_1 x + a_2 x^2$$
$$= a_1(x-1) + a_2(x^2 - 1), \text{ for any } a_1, a_2 \text{ real numbers.}$$

Therefore, $[(x-1)], (x^2 - 1)] = U$.

To check linear independence of the vectors $(x-1)$ and $(x^2 - 1)$, consider the linear combination of these two vectors, $\alpha(x-1) + \beta(x^2 - 1) = 0$

$$-\alpha - \beta + \alpha x + \beta x^2 = 0 \Rightarrow -\alpha - \beta = 0 \text{ and } \alpha = 0, \beta = 0.$$
$$\Rightarrow \alpha = \beta = 0. \text{ Hence } \{(x-1), (x^2 - 1)\} \text{ is LI.}$$

Therefore $\{(x-1), (x^2 - 1)\}$ is a basis of U. And dim $U = 2$.

Note: In the above example if $a_2 = -(a_0 + a_1)$, then the polynomial $p(x) = a_0 + a_1 x - (a_0 + a_1)x^2 = a_0(1 - x^2) + a_1(x - x^2)$, where a_0, a_1 varies over real numbers. As shown above, it can be checked that the subset

$$S = \{(1-x), (x - x^2)\} \text{ is also a basis for } U, \text{ dim } U = 2.$$

Example 40: Let $U = \{p(x) \in P_3 | \ p''(2) = 0\}$

Find a basis of U and hence dim U.

Let $p(x) = a_0 + a_1 x + a_2 x^2 + a_3 x^3$, for any scalars a_0, a_1, a_2, a_3.

$$p''(x) = 2a_2 + 6a_3 x$$

Since $p''(2) = 0 \Rightarrow 2a_2 + 12a_3 = 0$ i.e. $a_2 = -6a_3$.

$$U = \{p(x) = a_0 + a_1 x - 6a_3 x^2 + a_3 x^3\}$$
$$= \{a_0 + a_1 x - 6a_3 x^2 + a_3 x^3 | \text{ for any real numbers } a_0, a_1, a_3.\}$$
$$= [1, x, x^3 - 6x^2],$$

Therefore basis of $U = \{1, x, x^3 - 6x^2\}$, dim $U = 3$.

Example 41: Let $S = \{p(x) \in P_3 | p(1) + p'(1) = 0\}$. Find a basis of the subspace S.

Let $p(x) = a_0 + a_1 x + a_2 x^2 + a_3 x^3. \ p'(x) = a_1 + 2a_2 x + 3a_3 x^2.$

Therefore $p(1) + p'(1) = 0 \Rightarrow a_0 + a_1 + a_2 + a_3 + a_1 + 2a_2 + 3a_3 = 0$.

$a_0 + 2a_1 + 3a_2 + 4a_3 = 0 . \ a_0 = -2a_1 - 3a_2 - 4a_3.$

On substituting the values,

$$p(x) = -2a_1 - 3a_2 - 4a_3 + a_1 x + a_2 x^2 + a_3 x^3$$
$$= a_1(x - 2) + a_2(x^2 - 3) + a_3(x^3 - 4).$$

Therefore $S = [(x - 2), (x^2 - 3), (x^3 - 4)]$.

To check linear independence, consider the linear combination

$$\alpha(x - 2) + \beta(x^2 - 3) + \gamma(x^3 - 4) = 0.$$

On rearranging in powers of x, we get $-2\alpha - 3\beta - 4\gamma + \alpha x + \beta x^2 + \gamma x^3 = 0$.

On equating the coefficients $-\alpha - 3\beta - 4\gamma = 0$, $\alpha = 0$, $b = 0$, $\gamma = 0$.

Therefore set B is linearly independent.

Hence $B = \{(x - 2), (x^2 - 3), (x^3 - 4)\}$ is a basis of S. And dim $S = 3$.

Example 42: Extend the set $\{(1, 2, 3), (3, 2, 1)\}$ to a basis of V_3.

Obviously given set is L.I. Now to extend this to a basis of V_3, a third vector not a linear combination of these two vectors is to be introduced.

Consider vector (4, 4.4), which is sum of these two vectors, which a linear combination of (1, 1, 1). To make it independent, let us change one of the entries, say last i.e., so on taking (1, 1, 2). Now proposed basis is $B = \{(1, 2, 3), (3, 2, 1), (1, 1, 2)\}$.

To check the independence of the set B, consider $\alpha(1, 2, 3) + \beta(3, 2, 1) + \gamma(1, 1, 2) = (0, 0, 0)$.

On equating $\alpha + 3\beta + \gamma = 0$, $2\alpha + 2\beta + \gamma = 0$ and $3\alpha + \beta + 2\gamma = 0$.

On solving, we get, $\alpha = \beta = \gamma = 0$. Hence B is L.I.

So B is a basis of V_3.

Theorem 3.7: Let U and W be two subspaces of a N-dimensional vector space V.

Then dim $(U + W)$ = dim U + dim W − dim $(U \cap W)$.

We know that if U and W are subspaces of vector space V, then $(U \cap W)$ is also a subspace of V, but $U \cup W$ need not be a subspace; However $[U \cup W]$ is a subspace of V and $[U \cup W]$, where $U + W = \{v = u + w | u \in U$ and $w \in W\}$.

Let dim $U = m$, dim $W = n$ and dim $U \cap W = r$, where $m, n, r \leq N$.

Further Let $B = \{v_1, v_2, ..., v_r\}$ be a basis of $U \cap W$.

Extend B to a basis of U, by introducing some linearly independent vectors $\{u_1, u_2, ..., u_{m-r}\}$ to B i.e. so that $B_1 = \{v_1, v_2, ...v_r, u_1, u_2, ..., u_{m-r}\}$ becomes a basis of U. Similarly extend B to a basis of W by introducing linearly independent vectors $\{w_1, w_2, ..., w_{n-r}\}$ to B so that $B_2 = \{v_1, v_2, ..., v_r, w_1, w_2, ..., w_{n-r}\}$ becomes a basis of W.

Now consider set $B_3 = \{v_1, v_2, ..., v_r, u_1, u_2, ..., u_{m-r}, w_1, w_2, ..., w_{n-r}\}$ that B_3 spans $U + W$, which has $r + m - r + n - r = m + n - r$ vectors. Further if Linear independence of B_3 is proved then B_3 is a basis of $[U \cup W] = U + W$.

Now to prove linear independence, consider the combination

$\alpha_1 v_1 + ... + \alpha_r v_r + \beta_1 u_1 + ... + \beta_{m-r} u_{m-r} + \gamma_1 w_1 + ... + \gamma_{n-r} w_{n-r} = 0_{U+W}$.

On writing $\alpha_1 v_1 + ... + \alpha_r v_r = -(\beta_1 u_1 + ... + \beta_{m-r} u_{m-r} + \gamma_1 w_1 + ... + \gamma_{n-r} w_{n-r})$, we not that vector on left belongs to $(U \cap W)$ and vector on right side does not belong to $(U \cap W)$

Therefore above equality holds only when both sides are zero vectors i.e,

$\alpha_1 v_1 + ... + \alpha_r v_r = 0$, since set $\{v_1, v_2, ..., v_r\}$ is L.I. so each $\alpha_i = 0, 1, ... r$.

Now from other side $(\beta_1 u_1 + \ldots + \beta_{m-r} u_{m-r} + \gamma_1 w_1 + \ldots + \gamma_{n-r} w_{n-r}) = 0$, which can be written as $(\beta_1 u_1 + \ldots + \beta_{m-r} u_{m-r}) = -(\gamma_1 w_1 + \ldots + \gamma_{n-r} w_{n-r})$.

Vector on left belongs to U and vector on right belongs to W and none of these belongs to the intersection other than zero vector, therefore each is zero vector.

$(\beta_1 u_1 + \ldots + \beta_{m-r} u_{m-r}) = 0_U$, since set $\{v_1, v_2, \ldots, v_r, u_1, u_2, \ldots, u_{m-r}\}$ is L.I. $\beta_i = 0$, for $i = 1, \ldots m - r$

Similarly from $(\gamma_1 w_1 + \ldots + \gamma_{n-r} w_{n-r}) = 0$ all γ values are zeros.

Hence $B_3 = \{v_1, v_2, \ldots, v_r, u_1, u_2, \ldots, u_{m-r}, w_1, w_2, \ldots, w_{r-r}\}$ is L.I.

Hence $\dim [U \cup W] = \dim U + \dim W - \dim (U \cap W)$,

i.e., $\dim (U + W) = \dim U + \dim W - \dim (U \cap W)$.

Example 43: Let U and W be subspace of vector space V_3, defined by

$$U = \{(x_1, x_2, x_3) \in V_3 | x_1 + x_2 - 2x_3 = 0\} \text{ and}$$
$$W = \{(x_1, x_2, x_3) \in V_3 | x_1 - 3x_2 + 2x_3 = 0\}.$$

Find a basis of U and a basis of W and hence dimension of U and W. Then find a basis of $U \cap W$, and $\dim U \cap W$.

By definition $U = \left\{x_1, x_2, \dfrac{x_1 + x_2}{2}\right\} = \left\{\left(x_1, 0, \dfrac{x_1}{2}\right) + \left(0, x_2, \dfrac{x_2}{2}\right)\right\}$

$= \left\{\dfrac{x_1}{2}(2, 0, 1) + \dfrac{x_2}{2}(0, 2, 1)\right\} = \left[(2, 0, 1), (0, 2, 1)\right].$

Vectors $(2, 0, 1)$ and $(0, 2, 1)$ are linearly independent.

Hence a basis of $U = \{(2, 0, 1), (0, 2, 1)\}$ and $\dim U = 2$.

$W = \{(x_1, x_2, x_3) \in V_3 | x_1 - 3x_2 + 2x_3 = 0\}$

$= \left\{x_1, \dfrac{x_1 + 2x_3}{3}, x_3\right\}$, for x_1, x_3 any real numbers

$= \left\{\left(x_1, \dfrac{x_1}{3}, 0\right) + \left(0, \dfrac{2x_3}{3}, x_3\right)\right\}$

$= \left\{\dfrac{x_1}{3}(3, 1, 0) + \dfrac{x_3}{3}(0, 2, 3)\right\} = \left[(3, 1, 0), (0, 2, 3)\right].$

Two vectors $(3, 1, 0)$ and $(0, 2, 3)$ are linearly independent.

Hence $\{(3, 1, 0), (0, 2, 3)\}$ is a basis of W and $\dim W = 2$.

Vector Spaces

Now consider a subspace $S = U \cap W$ of V_3 i.e., defined by
$$S = \{(x_1, x_2, x_3) \in V_3 \mid x_1 + x_2 - 2x_3 = 0 \text{ and } x_1 - 3x_2 + 2x_3 = 0\}$$
On solving $x_1 + x_2 - 2x_3 = 0$ and $x_1 - 3x_2 + 2x_3 = 0$, we get $x_1 = x_2$, and $x_1 = x_3$.
$$S = \{(x_1, x_1, x_1) \mid \text{for any real number } x_1\}$$
$$= [(1, 1, 1)], \text{ since } (1, 1, 1) \text{ is non-zero vector, which is LI.}$$
Therefore $\{(1, 1, 1)\}$ is a basis of S, and dim $S = 1$,
Using dim $(U + W) = $ dim $U + $ dim $W - $ dim $(U \cap W) = 2 + 2 - 1 = 3$.
Therefore $U + W = V_3$.

Example 44: Let U and W be two subspaces of vector space V_3 defined by
$$U = \{(x_1, x_2, x_3) \in V_3 \mid 2x_1 - 3x_2 + 5x_3 = 0 \text{ and}$$
$$W = \{(x_1, x_2, x_3) \in V_3 \mid 4x_1 + x_2 - 3x_2 = 0.$$
Find bases of U, W and $U \cap W$ and dim U, dim W and dim $U \cap W$.
As given $U \cap W = \{v = (x_1, x_2, x_3) \in V_3 \mid 2x_1 - 3x_2 + 5x_3 = 0 \text{ and } 4x_1 + x_2 - 3x_3 = 0\}$.
$U = \{(x_1, x_2, x_3) \mid 2x_2 - 3x_2 + 5x_3 = 0\}$, on writing $3x_2 = 2x_1 + 5x_3$

$$= \left\{\left(x_1, \frac{2x_1 + 5x_3}{3}, x_3\right)\right\} = \left\{\frac{1}{3}(3x_1, 2x_1 + 5x_3, 3x_3)\right\}$$

$$= \left\{\frac{x_1}{3}(3, 2, 0) + \frac{x_3}{3}(0, 5, 3)\right\}$$

$$= [(3, 2, 0), (0, 5, 3)] \text{ since } x_1 \text{ and } x_3 \text{ are any real numbers.}$$

Obviously $(3, 2, 0)$ and $(0, 5, 3)$ are linearly independent.
Hence $\{(3, 2, 0), (0, 5, 3)\}$ is a basis of U, dim $U = 2$.
Now consider
$$W = \{(x_1, x_2, x_3) \mid 4x_1 + x_2 - 3x_3 = 0\}$$

$$= \left\{\left(x_1, x_2, \frac{4x_1 + x_2}{3}\right)\right\}, \text{ from } 3x_1 + x_2 - 3x_3 = 0$$

$$= \left\{\frac{1}{3}(3x_1, 3x_2, 4x_1 + x_2)\right\}$$

$$= \left\{\frac{x_1}{3}(3, 0, 4) + \frac{x_2}{3}(0, 3, 1)\right\}$$

$= [(3, 0, 4), (0, 3, 1)]$, for any real numbers x_1, x_2

This shows $[(3, 0, 4), (0, 3, 1)] = W$, and $(3, 0, 4)$ and $(0, 3, 1)$ are LI.

Therefore dim $W = 2$.

Now to find a basis of the intersection of U and W, considering

$$U \cap W = \{(x_1, x_2, x_3) \mid 2x_1 - 3x_2 + 5x_3 = 0 \text{ and } 4x_1 + x_2 - 3x_3 = 0\}$$

$$= \left\{\left(x_1, \frac{13}{2}x_1, \frac{7}{2}x_1\right), \text{ on solving for } x_2 \text{ and } x_3\right\}$$

$$= \left\{\frac{x_1}{2}(2, 13, 7)\right\} = [(2, 13, 7)], \text{ span of a single vector } (2, 13, 7).$$

This is non zero vector, which is linearly independent.

Therefore   dim $U \cap W = 1$

Hence   dim $(U + W) = $ dim $U + $ dim $W - $ dim $(U \cap W)$

$$= 2 + 2 - 1 = 3.$$

This shows that $U + W$, since dim $(U + W) = 3$.

Now to find a basis of $(U + W)$, one more linearly independent vector is to be added to the basis of U or W.

To find such a vector, on taking $(3, 2, 0) + (0, 5, 3) = (3, 7, 3)$ i.e. $(3, 7, 3)$ is linear combination of other two vectors.

Now changing to $(3, 6, 3)$ i.e. $3(1, 2, 1)$ to get third linearly independent vector.

Further check $(3, 2, 0), (0, 5, 3), (1, 2, 1)$ is LI, which is extended basis of V_3, obtained by extending the basis of U.

Similarly $\{(3, 0, 4), (0, 3, 1), (1, 2, 1)\}$ is L.I. This can be shown that this set is also basis of V_3 by introducing one vector $(1, 2, 1)$ to the basis of W.

Hence bases of U and W have been extended to two bases of V_3 by introducing one more vector to bases of U and W respectively.

Example 45: Determine a basis and dimension of (a) $[S_1] + [S_2]$ and (b) $[S_1] \cap [S_2]$ for two given set $S_1 = \{(1, 2, 3), (0, 1, 2), (3, 2, 1)\}$ and

$$S_2 = \{(1, -2, 3), (-1, 1, -2), (1, -3, 4)\} \text{ of } V_3$$

On checking $S_1 = \{1, 2, 3) (0, 1, 2) (3, 2, 1)\}$ is found linearly dependent and a linearly independent subset of S_1 is $A_1 = \{(1, 2, 3), (0, 1, 2)\}$ such that $[S_1] = [A_1]$.

$S_2 = \{(1, -2, 3), (-1, 1, -2), (1, -3, 4)\}$ is also linearly dependent, and a linearly independent subset $A_2 = \{(1, -2, 3) (-1, 1, -2)\}$ is such that $[A_2] = [S_2]$.

Therefore dim $[S_1] = 2$, dim $[S_2] = 2$ and basis of $[S_1] + [S_2]$ has 3 linearly independent vectors, because 4 vectors $\{(1, 2, 3), (0, 1, 2), (1, -2, 3), (-1, 1, -2)\}$ in V_3 is linearly dependent and $\{(1, 2, 3), (0, 1, 2), (1, -2, 3)\}$ is linearly independent.

Hence dim $([S_1] + [S_2]) = 3$.

Now using dim $([S_1] + [S_2] = \dim S_1] + \dim [S_2] - \dim [S_1] \cap [S_2]$,

We get $3 = 2 + 2 - \dim [S_1] \cap [S_2]$ i.e., dim $[S_1] \cap [S_2] = 1$.

Now to find a basis of $[S_1] \cap [S_2]$, we have to find a vector which is in the intersection of $[S_1]$ and $[S_2]$.

To find such a vector consider. $\alpha(1, 2, 3) + \beta(0, 1, 2) = \gamma(1, -2, 3) + \delta(-1, 1, -2)$.

$$(\alpha, 2\alpha, 3\alpha) + (0, \beta, 2\beta) = (\gamma, -2\gamma, 3\gamma) + (-\delta, \delta, -2\delta)$$

$\Rightarrow$ $\quad (\alpha + 0, 2\alpha + \beta, 3\alpha + 2\beta) = (\gamma - \delta, -2\gamma + \delta, 3\gamma - 2\delta)$.

On equating

$$\alpha + 0 = \gamma - \delta, \; 2\alpha + \beta = -2\gamma + \delta, \; 3\alpha + 2\beta = 3\gamma - 2\delta$$

$\Rightarrow$ $\quad \alpha + 0\beta - \gamma + \delta = 0, \; 2\alpha + \beta + 2\gamma - \delta = 0, \; 3\alpha + 2\beta - 3\gamma + 2\delta = 0$.

Augmented matrix for these equations is:

$$\begin{bmatrix} 1 & 0 & -1 & 1 & 0 \\ 2 & 1 & 2 & -1 & 0 \\ 3 & 2 & -3 & 2 & 0 \end{bmatrix} \sim \begin{bmatrix} 1 & 0 & -1 & 1 & 0 \\ 0 & 1 & 4 & -3 & 0 \\ 0 & 2 & 0 & -1 & 0 \end{bmatrix} \sim \begin{bmatrix} 1 & 0 & -1 & 1 & 0 \\ 0 & 1 & 4 & -3 & 0 \\ 0 & 0 & -8 & 5 & 0 \end{bmatrix}$$

$\Rightarrow$ $\quad \alpha - \gamma + \delta = 0, \; \beta + 4\gamma - 3\delta = 0, \; -8\gamma + 5\delta = 0$

On choosing $\delta = 8$, $\gamma = 5$, $\alpha = -3$ and $\beta = 4$. Common vector is $(-3, 10, 17)$.

Therefore a basis of $[S1] \cap [S2] = \{(3, -10, -17)\}$.

Definition 2.9: Let $B = \{u_1, u_2, ..., u_n\}$ be an ordered basis for an n-dimensional vector space V. Further let $U \in V$ where $u = \alpha_1 u_1 + \alpha_2 u_2 + , ..., + \alpha_n u_n$ scalars $(\alpha_1, \alpha_2, ..., \alpha_n)$ are called coordinates of vector u with respect to the ordered basis B.

Example 46: Find the coordinates of the vector $(3, 4, 5)$ with respect to the ordered basis $B = \{(1, 0, 1), (1, 1, 0), (0, 1, 1)\}$ of R^3.

Let
$$(3, 4, 5) = \alpha(1, 0, 1) + \beta(1, 1, 0) + \gamma(0, 1, 1)$$
$$(3, 4, 5) = (\alpha + \beta, \beta + \gamma, \alpha + \gamma)$$

$\Rightarrow$ $\quad \alpha + \beta = 3, \; \beta + \gamma = 4, \; \alpha + \gamma = 5.$

On solving $\alpha = 2$, $\beta = 1$ and $\gamma = 4$.

Therefore $(2, 1, 4)$ are coordinates of the vector $(3, 4, 5)$ with respect to given basis B.

Example 47: Find coordinates of the vector $(3, -2, 5, -4)$ with respect to the ordered basis $\{(0, 1, 1, 1), (1, 0, 1, 1), (1, 1, 0, 1), (1, 1, 1, 0)\}$.

Consider linear combination

$$\alpha(0, 1, 1, 1) + \beta(1, 0, 1, 1) + \gamma(1, 1, 0, 1) + \delta(1, 1, 1, 0) = (3, -2, 5, -4)$$

On simplifying, we get $(\beta + \gamma + \delta, \alpha + \gamma + \delta, \alpha + \beta + \delta, \alpha + \beta + \gamma) = (3, -2, 5, -4)$,

On equating, we get $\beta + \gamma + \delta = 3$, $\alpha + \gamma + \delta = -2$, $\alpha + \beta + \delta = 5$, $\alpha + \beta + \gamma = -4$.

On adding all, we get $\alpha + \beta + \gamma + \delta = 0$, giving $\alpha = -3$, $\beta = 2$, $\gamma = -5$, $\delta = 4$.

Hence the coordinates of $(3, -2, 5, -4)$ with respect to the given ordered basis is $(-3, 2, -5, 4)$.

Example 48: Find the coordinates of $x^2 + 2x - 1$ with respect to ordered basis

$$B = \{x + 1, x^2 + x - 1, x^2 - x + 1\} \text{ of } P_2.$$

Let $\quad x^2 + 2x - 1 = \alpha(x + 1) + \beta(x^2 + x - 1) + \gamma(x^2 - x + 1)$

$\qquad = (\beta + \gamma)x^2 + (\alpha + \beta - \gamma)x + \alpha - \beta + \gamma$

Equating the coefficients of powers of x, we get

$$\beta + \gamma = 1, \quad \alpha + \beta - \gamma = 2, \quad \alpha - \beta + \gamma = -1.$$

On solving these equations, we get $2\beta - 2\gamma = 3$, $2\beta + 2\gamma = 2$, $4\beta = 5$,

$$\beta = \frac{5}{4}, \gamma = -\frac{1}{4} \Rightarrow \alpha = \frac{1}{2}, \beta = \frac{5}{4} \text{ and } \gamma = -\frac{1}{4}.$$

Therefore $(\alpha, \beta, \gamma) = \left(\frac{1}{2}, \frac{5}{4}, -\frac{1}{4}\right)$ is required coordinates.

Example 49: Find the coordinates of $2 - 3x + 4x^2 - x^3$ with respect to the ordered basis $\{1, x, x^2, x^3 - 1\}$ of P_3

Let $\alpha + \beta x + \gamma x^2 + \delta(x^3 - 1) = 2 - 3x + 4x^2 - x^3$, on rearranging, we get

$$(\alpha - \delta) + \beta x + \gamma x^2 + \delta x^3 = 2 - 3x + 4x^2 - x^3.$$

Equating coefficients of powers of x we get,

$$\alpha - \delta = 2, \beta = -3, \gamma = 4, \delta = -1, \Rightarrow \alpha = 1,$$

$(\alpha, \beta, \gamma, \delta) = (1, -3, 4, -1)$ are coordinates of the given polynomial with respect to the given basis.

Vector Spaces

EXERCISE SET 3

1. Prove that set R of all real numbers is a vector space with usual addition and multiplication of real numbers

2. Prove that set $S = \{\{x, 0) | x$ is any real number$\}$ is a vector space with rule of addition as $(x_1, 0) + (x_2, 0) = (x_1 + x_2, 0)$ and scalar multiplication $\alpha(x, 0) = (\alpha x, 0)$.

3. Let P_n be the space of polynomials of degree $\leq n$. Prove that
$S = \{p(x) \in P_n | p(x) = x\, p'(x)\}$ is a vector space.

4. Let P_n be the space of polynomials of degree $\leq n$. Prove that
$S = \{p(x) \in P_n | p(x) = (x^2 + x)\, p^2(x)\}$ is a vector space.

5. Let P_n be the space of polynomials of degree $\leq n$.
Let $U = \{p(x) \in P_n | p'(1) = p''(2)\}$. Prove that U is a subspace of P_n.

6. Prove that $U = \{(x_1, x_2, x_3, x_4, x_5) \in R^5 | x_1 + x_5 = 0\}$ is a vector space with usual addition of vectors and scalar multiplication.

7. Let V be a subset of R^4 such that $2x_1 - 3x_2 + 4x_3 - 5x_4 = 0$ for all $(x_1, x_2, x_3, x_4) \in R^4$ Prove that V is a subspace of R^4.

8. Let V be a subset of R^4 such that $x_1^2 + x_2^2 + x_3^2 + x_4^2 = 0$ for $(x_1, x_2, x_3, x_4) \in R^4$. Prove that V is a subspace of R^4.

9. $U = \{f(x) \in C[-1, 1] | f''(0) = 0\}$. Prove that U is a vector space with point-wise addition and scalar multiplication.

10. Prove that $p_n(x)$ space of polynomials of degree $\leq n$ is a subspace of $C[-\infty, \infty]$ with point-wise addition and scalar multiplication.

11. Let $S = \{f(x) \in C[a, b] | f(1) + f'(-1) = f(0)\}$. Prove that S is a subspace of $C[a, b]$.

12. Let P_5 be the space of all polynomials of degree ≤ 5, with usual rule of polynomial addition and scalar multiplication. Prove that the set $U = \{p(x) \in P_5\ p''(-1) = 0\}$ is a subspace of P_5.

13. Let $U = \left\{f(x) \in C[a, b] | 2f\left(f\left(\dfrac{a+b}{2}\right)\right) = f(a) + f(b)\right\}$. Prove that U is a vector space with point-wise addition and scalar multiplication.

14. Let $U = [(1, 0, 0)]$ and $W = (0, 1, 0)]$ be subspaces of R^3. Show that $U \cup W$ is not a subspace of R^3.

15. Find the value of m, such that the vector $(m, 7, -4)$ is a linear combination of vectors $(-2, 2, 1)$ and $(2, 1, -2)$.

16. Find a such that set $S = \{(2, -1, 3), (3, 4, -1), (a, 2, 1)\}$ is linearly independent.
17. Check whether the set $S = \{(4, -6, 8), (-5, 2, 3), (-3, -1, 7)\}$ is linearly independent.
18. Check whether the set $S = \{2, x-1, x+x^2, x^2+x^3, x^3+1\}$ is linearly independent.
19. Check whether the set $S = \{(x-1), (x-1)^2, (x-1)^3, 1\}$ is linearly independent. Further show that $[S] = P_3$.
20. Let $\{v_1, v_2, v_3\}$ be a linearly independent subset of a vector space V. Prove or disprove, if $w_1 = v_1 + 2v_2 + v_3$, $w_2 = 2v_1 + v_2 + v_3$ and $w_3 = v_1 + v_2 + 2v_3$ then set $\{w_1, w_2, w_3\}$ is L.I.
21. Let $\{v_1, v_2, v_3\}$ be a linearly independent subset of a vector space V. Prove or disprove, if $w_1 = v_1 - 2v_2 + v_3$, $w_2 = -2v_1 + v_2 + v_3$ and $w_3 = v_1 + v_2 - 2v_3$ then set $\{w_1, w_2, w_3\}$ is L.D.
23. Let $S = \{u_1, u_2, ..., u_n\}$ be a linearly independent set in a vector space V_n. Check whether the set $S = \{u_1 + u_2, u_2 + u_3, u_3 + u_4, ..., u_{n-1} + u_n, u_n + u_1\}$ linearly dependent or linearly independent in V_n, with n odd.
24. If the set $S = \{u, v, w\}$ is a basis of V_3, then show that the set
$$S_1 = \{u + v, v + w, w + u\} \text{ is also a basis for } V_3.$$
25. Find the number linearly independent vectors of the set
$$S = \{(3, 4, 1, -5), (1, -2, 3, 4), (4, 2, 4, -1), (2, 6, -2, -9)\}.$$
26. Find a basis of the space spanned by the set $\{1 + x, x + x^2, x^2 - 2, 3 - 4x + 5x^2\}$ by using row reduced echelon method.
27. Let $U = \{(x_1, x_2, x_3) | x_1 - x_2 + x_3 = 0\}$ be a subspace of V_3. Find a basis of U and hence find dim U.
28. Let $U = \{(x_1, x_2, x_3, x_4) | x_1 - 2x_2 + x_3 - 3x_4 = 0\}$ be a subspace of V_3. Find a basis of U and hence find dim U.
29. Find a basis of subspace $U = \{p(x) \in P_5 | p''(-1) = 0\}$ of P_5, and hence dim U.
30. Show that set $U = \{(x, y, 2x - 3y) | x, y \text{ are real numbers}\}$ is a subspace of V_3. Find a basis of U and hence find dim U.
31. Show that set $U = \{(x - y, y - 2z, x + 3z) | x, y, z \text{ are real numbers}\}$ is a subspace of V_3. Find a basis of U and hence find dim U.
32. Let $S = \{2, x, x - x^2, x + x^2\}$ be a subset of p_2. Find the dimension of $[S]$.
33. Let $S_1 = \{(1, 2, 0, 0), (4, 7, 5, 0)\}$ and $S_2 = \{(1, 1, 0, 0), (1, 0, 4, 0)\}$ be two subsets of V_4. Find the dimension of $[S_1] \cap [S_2]$.
34. Let $S = \{2 - 2x, -2 + 2x + x^2, 2 - 2x + x^2\}$ be a subset of p_2.
Check whether S is a basis for p_2, if not determine dim $[S]$.

35. Let $S = \{(2, 3, 1), (1, -2, 3), (3, 1, 4), (4, -2, 5)\}$ be subset of V_3, find a linearly independent subset A of S, such that $[A] = [S]$.

36. Let $S = \{(1, -3, 7), (2, 0, -6), (3, -1, -1), (2, 4, -5)\}$ be subset of V_3, find a linearly independent subset A of S, such that $[A] = [S]$.

37. Find the bases and dimensions of the following subspaces of p_3;
    (a) $U = \{p(x) \in P_3 | \; p(1) = p'(1) = 0\}$
    (b) $W = \{p(x) \in P_3 | \; p(0) = p'(0) = p''(0) = 0\}$

38. Let $\{u, v, w\}$ be a basis of V. Is subset $B = [u + v - w, u - v + w, -u + v + w]$ a basis for V. What is dim $[B]$?

39. Set $S = \{(3, 2, 0, 1), (1, 0, 5, 3), (0, 1, 2, 5), (2, -1, 3, 0)\}$ be a set in V_4. Is $[S] = V_4$. Find dim $[S]$.

40. Let $U = \{(x_1, x_2, x_3, x_4, x_5)\} \in V_5 |, x_1 - x_2 - x_3 = 0$, and $2x_1 - x_4 + 2x_5 = 0\}$. Find a basis of U, and hence dim U.

41. Find the coordinates of vector $(-1, 2, 3)$ relative to the ordered basis
    $$B = \{(1, 0, 0), (0, 1, 0), (0, 0, 1)\} \text{ of } V_3.$$

42. Find the coordinates of $x^2 + 2x - 7$ relative to the ordered basis $\{1, x, x^2\}$ of P_2.

43. Find the coordinates of vector $(-1, 2, 3)$ relative to the ordered basis
    $$B = \{(1, 1, 1), (2, 0, 1), (2, 3, 5)\} \text{ of } V_3.$$

44. Find the coordinates of $1 + 2x + x^2$ relative to the ordered basis
    $$\{x - 2, x^2 + x, 1 + x^2\} \text{ of } P_2.$$

45. Find the coordinates of $x^3 + 3x^2 - x + 4$ relative to the ordered basis
    $$\{1, x - 1, x + x^2, x^2 + x^3\} \text{ of } P_3.$$

ANSWERS TO EXERCISE SET – 3

15. $m = 2$.   16. $a \neq 3$.   17. L.D.   18. LD   19. LI   20. LI   21. LD   23. LD
25. two
26. $\{1 + x, x + x^2, x^2 - 2\}$,   27. $\{(1, 1, 0), (0, 1, 1)\}$ dim $U = 2$,   28. dim $U = 3$,
29. Basis $= \{1, x, x^3 + 3x^2, x^4 - 6x^2, x^5 + 10x^2\}$, dim $U = 5$
30. Basis $= \{(1, 0, 2), (0, 1, -3)\}$, dim $U = 2$.
31. Basis $= \{(1, 0, 1), (-1, 1, 0), (0, -2, 3)\}$, dim $U = 3$
32. dim$[S] = 3$, 33. 1, 34. dim$[S] = 2$, 35. $A = \{(2, 3, 1), (1, -2, 3), (4, -2, 5)\}$

36. Any three vectors including first vector.
37. (a) $B = \{1 - 2x + x^2, 2 - 3x + x^3\}$, dim = 2
    (b) Basis of $W = \{x^3\}$, dim $W = 1$, 38. Yes, dim[B] = 3, 39. Yes dim[B] = 4,
40. Basis $B = \{1, 1, 0, 2, 0), (0, -1, 1, 0, 0), (0, 0, 0, 2, 1)\}$, dim $U = 3$
41. Coordinates $(-1, 2, 3)$, 42. Coordinates $(-7, 2, 1)$, 43. Coordinates $(-1, -1, 1)$,
44. Coordinates $(-2, 4, -3)$, 45. Coordinates $(1, -3, 2, 1)$.

CHAPTER 4

Linear Transformations

INTRODUCTION

In many situations, where vector spaces and related subject are being discussed and used, problems and their solutions are not obvious or are difficult to analysis them. In such cases, if the vector spaces are transformed into another vector space by a suitable transformation, by keeping certain basic properties of the problem, the analysis of the solution of the problem becomes easier.

For example, in R^2, two vectors $u = 3\vec{i} - 4\vec{j}$ and $v = 4\vec{i} + 3\vec{j}$ can be transformed to $u_0 = \vec{i}$ and $v_0 = \vec{j}$ respectively, if we apply matrix multiplication by

$$\begin{bmatrix} \dfrac{3}{25} & \dfrac{4}{25} \\ -\dfrac{4}{25} & \dfrac{3}{25} \end{bmatrix} \begin{bmatrix} 3\vec{i} - 4\vec{j} \\ 4\vec{i} + 3\vec{j} \end{bmatrix} = \begin{bmatrix} \vec{i} \\ \vec{j} \end{bmatrix}.$$

If $\begin{bmatrix} \dfrac{3}{25} & \dfrac{4}{25} \\ -\dfrac{4}{25} & \dfrac{3}{25} \end{bmatrix}$ denoted by T, then equivalently T is a map from $R^2 \to R^2$ denoted by $T: R^2 \to R^2$, such that $T(3, 4) = (1, 0)$ and $T(-4, 3) = (0, 1)$.

Since, $\begin{bmatrix} \dfrac{3}{25} & \dfrac{4}{25} \\ -\dfrac{4}{25} & \dfrac{3}{25} \end{bmatrix} \begin{bmatrix} 3 \\ 4 \end{bmatrix} = \begin{bmatrix} 1 \\ 0 \end{bmatrix}, \begin{bmatrix} \dfrac{3}{25} & \dfrac{4}{25} \\ -\dfrac{4}{25} & \dfrac{3}{25} \end{bmatrix} \begin{bmatrix} -4 \\ 3 \end{bmatrix} = \begin{bmatrix} 0 \\ 1 \end{bmatrix}.$

Now such transformations are defined below and are discussed in this chapter.

4.1 LINEAR TRANSFORMATIONS

The following known transformations $R^2 \to R^2$
- (i) Projection on x-axis $\quad T(x, y) = (x, 0)$
- (ii) Projection on y-axis $\quad T(x, y) = (0, y)$
- (iii) Reflection about x-axis $\quad T(x, y) = (x, -y)$
- (iv) Reflection about y-axis $\quad T(x, y) = (-x, y)$
- (v) Rotation about origin by angle θ.

$(x, y) \to (x', y')$, where $x' = x \cos \theta + y \sin \theta$, $y' = -x \sin \theta + y \cos \theta$, satisfying certain conditions are linear.

The formal definition of linear transformation is given below:

Definition 4.1: Let U and V be two vector spaces over the same set of scalars.

A map $T : U \to V$ from U to V is called a linear map, or linear transformation, if T satisfies the following two axioms.

(1) $T(u_1 + u_2) = T(u_1) + T(u_2)$ for all $u_1, u_2 \in U$, and $T(u_1), T(u_2) \in V$. Addition of u_1, u_2 on left is addition of vectors of U and addition of $T(u_1), T(u_2)$ on right is addition of vector space V.

(2) $T(\alpha u_1) = \alpha T(u_1)$ for all $u_1 \in U$ and all scalars α scalar multiplication on left is of vector space U and on right $\alpha T(u_1)$ is scalar multiplication of V.

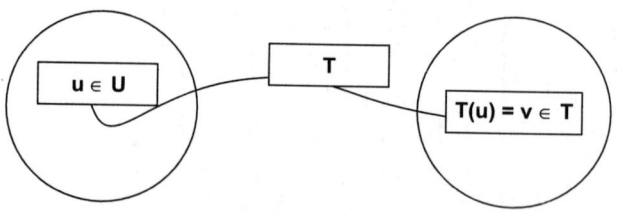

Example 1: Let $U = V_3$ and $V = V_2$, and $T : V_3 \to V_2$ be defined by $T(x_1, x_2, x_3) = (x_1 - x_2, x_2 + x_3)$ for all $\forall (x_1, x_2, x_3) \in V_3$.

Let $u_1 = (x_1, x_2, x_3)$ and $u_2 = (y_1, y_2, y_3)$ be any two vectors of V_3.

Addition $\quad u_1 + u_2 = (x_1 + y_1, x_2 + y_2, x_3 + y_3)$

$$T(u_1 + u_2) = T(x_1 + y_1, x_2 + y_2, x_3 + y_3)$$

$$= (x_1 + y_1 - x_2 - y_2, x_2 + y_2 + x_3 + y_3)$$
$$= (x_1 - x_2, x_2 + x_3) + (y_1 - y_2, y_2 + y_3)$$
by definition of transformation
$$= T(x_1, x_2, x_3) + T(y_1, y_2, y_3)$$
$$= T(u_1) + T(u_2). \; \forall \; u_1, u_2 \in U.$$

For any scalar α and vector $u_1 \in V_3$, $\alpha u_1 = \alpha(x_1, x_2, x_3) = (\alpha x_1, \alpha x_2, \alpha x_3)$.

now
$$T(\alpha u_1) = T(\alpha x_1, \alpha x_2, \alpha x_3) = (\alpha x_1 - \alpha x_2, \alpha x_2 + \alpha x_3)$$
$$= \alpha(x_1 - x_2, x_2 + x_3) = \alpha T(x_1, x_2, x_3)$$
$$= \alpha T(u_1) \text{ for all scalar } \alpha \text{ and } u_1 \in U.$$

Therefore T is a linear map.

Example 2: Let $T : P_2 \to V_3$ be defined by $T(p(x) = \alpha + \beta x + \gamma x^2 \in P_2) = (\alpha, \beta, \gamma)$. Check T is a linear map.

Let $p(x) = \alpha + \beta_1 x + \gamma x^2$ and $q(x) = \alpha_1 + \beta_1 x + \gamma_1 x^2 \in P_2$

Now $p(x) + q(x) = (\alpha + \alpha_1) + (\beta + \beta_1)x + (\gamma + \gamma_1)x^2$ by addition of polynomials.

$$T(p(x) + q(x)) = T((\alpha + \alpha_1) + (\beta + \beta_1)x + (\gamma + \gamma_1)x^2)$$
$$= (\alpha + \alpha_1, \beta + \beta_1, \gamma + \gamma_1) \text{ by definition}$$
$$= (\alpha, \beta, \gamma) + (\alpha_1, \beta_1, \gamma_1) \text{ by addition of addition}$$
$$= T(\alpha + \beta x + \gamma x^2) + T(\alpha_1 + \beta_1 x + \gamma_1 x^2)$$
$$= Tp(x) + T(q(x)) \; \forall \; p(x), q(x) \in P_2$$

Also $T(\alpha' p(x)) = T(\alpha'\alpha + \alpha'\beta x + \alpha'\gamma x^2) = (\alpha'\alpha, \alpha'\beta, \alpha'\gamma)$
$$= \alpha'(\alpha, \beta, \gamma) = \alpha' T(\alpha + \beta x + \gamma x^2) = \alpha' Tp(x)$$
for all scalars α'.

Therefore T is a linear map.

Example 3: Let $T: P_2 \to V_3$ be a map defined by
$$T(a_0 + a_1 x + a_2 x^2) = (a_2 - a_0, a_0 + a_1, a_1 + a_2).$$
Check whether T is linear.

Let $p(x) = a_0 + a_1 x + a_2 x^2$, $q(x) = b_0 + b_1 x + b_2 x^2 \in P_2$.
$$p(x) + q(x) = (a_0 + b_0) + (a_1 + b_1)x + (a_2 + b_2)x^2$$
$$\begin{aligned}
T(p(x) + q(x)) &= T((a_0 + b_0) + (a_1 + b_1)x + (a_2 + b_2)x^2) \\
&= (a_2 + b_2 - a_0 - b_0, a_0 + b_0 + a_1 + b_1, a_1 + b_1 + a_2 + b_2) \\
&= (a_2 - a_0, a_0 + a_1, a_1 + a_2) + (b_2 - b_0, b_0 + b_1, b_1 + b_2) \\
&= T(a_0 + a_1 x + a_2 x^2) + T(b_0 + b_1 x + b_2 x^2) \\
&= T(p(x)) + T(q(x))
\end{aligned}$$

$$\begin{aligned}
T(\alpha p(x)) &= T(\alpha a_0 + \alpha a_1 x + \alpha a_2 x^2) \\
&= (\alpha a_2 - \alpha a_0, \alpha a_0 + \alpha a_1, \alpha a_1 + \alpha a_2) \\
&= \alpha(a_2 - a_0, a_0 + a_1, a_1 + a_2) \\
&= \alpha\, T(p(x)), \text{ for all scalars } \alpha.
\end{aligned}$$

Hence T is a linear map.

Example 4: Let $T: V_2 \to V_3$ be a map defined by $T(x_1, x_2) = (x_1, x_1 x_2, x_2)$. Check whether T is linear.

Let $(x_1, x_2), (y_1, y_2) \in V_2$ then
$T(x_1, x_2) = (x_1, x_1 x_2, x_2)$ and $T(y_1, y_2) = (y_1, y_1 y_2, y_2)$.
$$\begin{aligned}
T((x_1, x_2) + (y_1, y_2)) &= T(x_1 + y_1, x_2 + y_2) \\
&= (x_1 + y_1, (x_1 + y_1)(x_2 + y_2), x_2 + y_2) \\
&= (x_1 + y_1, x_1 x_2 + y_1 y_2 + x_1 y_2 + y_1 x_2, x_2 + y_2) \\
&= (x_1, x_1 x_2, x_2) + (y_1, y_1 y_2, y_2) + (0, x_1 y_2 + y_1 x_2, 0) \\
&= T(x_1, x_2) + T(y_1, y_2) + (0, x_1 y_2 + y_1 x_2, 0) \\
&\neq T(x_1, x_2) + T(y_1, y_2) \text{ for all } (x_1, x_2), (y_1, y_2) \in V_2.
\end{aligned}$$

Hence T is not a linear map.

Also $T(\alpha(x_1, x_2)) = T(\alpha x_1, \alpha x_2) = (\alpha x_1, \alpha x_1 \alpha x_2, \alpha x_2)$
$$= (\alpha x_1, \alpha x_1 \alpha x_2, \alpha x_2) = \alpha(x_1, \alpha^2 x_1 x_2, x_2)$$
$$\neq \alpha T(x_1, x_2).$$

Hence T is not linear.

If one of the above two conditions is not satisfied then T is not linear map.

Example 5: Let $T: P_3 \to P_2$ be defined by $T(\alpha + \beta x + \gamma x^2 + \delta x^3) = (\alpha + \beta) + (\beta + \gamma)x + (\gamma + \delta)x^2$. Show that T is linear.

Let $p(x) = \alpha_1 + \beta_1 x + \gamma_1 x^2 + \delta_1 x^3$ and $q(x) = \alpha_2 + \beta_2 x + \gamma_2 x^2 + \delta_2 x^3$.

$$\begin{aligned}
T(p(x) + q(x)) &= T(\alpha_1 + \beta_1 x + \gamma_1 x^2 + \delta_1 x^3 + \alpha_2 + \beta_2 x + \gamma_2 x^2 + \delta_2 x^3) \\
&= T((\alpha_1 + \alpha_2 + \beta_1 x + \beta_2 x + \gamma_1 x^2 + \gamma_2 x^2 + \delta_1 x^3 + \delta_2 x^3) \\
&= T((\alpha_1 + \alpha_2) + (\beta_1 + \beta_2)x + (\gamma_1 + \gamma_2)x^2 + (\delta_1 + \delta_2)x^3) \\
&= (\alpha_1 + \alpha_2 + \beta_1 + \beta_2) + (\beta_1 + \beta_2 + \gamma_1 + \gamma_2)x + (\gamma_1 + \gamma_2 + \delta_1 + \delta_2)x^2 \\
&= (\alpha_1 + \beta_1) + (\beta_1 + \gamma_1)x + (\gamma_1 + \delta_1)x^2 + (\alpha_2 + \beta_2) + (\beta_2 + \gamma_2)x \\
&\quad + (\gamma_2 + \delta_2)x^2 \\
&= Tp(x) + Tq(x)) = T(p(x) + q(x)).
\end{aligned}$$

Now $\alpha p(x) = \alpha(\alpha_1 + \beta_1 x + \gamma_1 x^2 + \delta_1 x^3) = \alpha\alpha_1 + \alpha\beta_1 x + \alpha\gamma_1 x^2 + \alpha\delta_1 x^3$.

$$\begin{aligned}
T(\alpha p(x)) &= T(\alpha\alpha_1 + \alpha\beta_1 x + \alpha\gamma_1 x^2 + \alpha\delta_1 x^3) \\
&= (\alpha\alpha_1 + \alpha\beta_1) + (\alpha\beta_1 + \alpha\gamma_1)x + (\alpha\gamma_1 + \alpha\delta_1)x^2 \\
&= \alpha((\alpha_1 + \beta_1) + (\beta_1 + \gamma_1)x + (\gamma_1 + \delta_1)x^2) \\
&= \alpha T(p(x)), \text{ for every scalar } \alpha. \text{ Hence } T \text{ is a linear map.}
\end{aligned}$$

Example 6: Let $C[-1, 1]$ be the vector space of all continuous functions on the interval $[-1, 1]$, a map T is defined as : $T(f(x)) = x f'(x)$ for all $f(x) \in C[-1, 1]$.

Show that T is a linear map.

Let $f(x), g(x) \in C[-1, 1]$.
$$\begin{aligned}
T(f(x) + g(x)) &= x(f + g)'(x) \\
&= x(f'(x) + g'(x)) \\
&= x f'(x) + x g'(x) \\
&= T f(x) + T g(x).
\end{aligned}$$

$$T(af(x)) = x[af(x)]' = x(af'(x)) = aTf(x).$$

Therefore T is a linear map.

Theorem 4.1: Let $T: U \to V$ be a linear map then

(a) $T(0_u) = 0_v$

(b) $T(-u) = -T(u), \forall u \in U$

(c) $T(\alpha_1 u_1 + \alpha_2 u_2 + ,..., \alpha_n u_n) = \alpha_1 T(u_1) + ,..., + \alpha_n T(u_n)$

4.6   Elementary Linear Algebra

for all $u_1, u_2, \ldots, u_n \in U$ and any scalars $\alpha_1, \alpha_2, \ldots, \alpha_n$.

In other words, a linear map T transforms the zero vector of vector space U into the zero vector of vector space V and additive inverse of every vector $u \in U$ into the additive inverse of $T(u)$ in V.

Proof:

(a) Choosing $\alpha = 0$ in $T(\alpha u) = \alpha T(u)$ because T is linear, $T(0u) = 0T(u) \Rightarrow T(0_u) = 0_v$, since $Tu \in V$.

(b) Choose $\alpha = -1$, $T(-1u) = -1T(u) \Rightarrow T(-u) = -T(u)$

(c) Using $T(\alpha_1 u_1 + \alpha_2 u_2) = \alpha_1 T(u_1) + \alpha_2 T(u_2)$, since this is equivalent to conditions of T linear.

$$T(\alpha_1 u_1 + \alpha_2 u_2 + \ldots + \alpha_n u_n) = T(\alpha_1 u_1) + T(\alpha_2 u_2 + \ldots + \alpha_n u_n)$$

$$= \alpha_1 T(u_1) + T(\alpha_2 u_2) + T(\alpha_3 u_3 + \ldots + \alpha_n u_n), \text{ since } T \text{ is linear}$$

$$= \alpha_1 T(u_1) + \alpha_2 T(u_2) + T(\alpha_3 u_3) + T(\alpha_4 u_4 + \ldots + \alpha_n u_n).$$

Continuing similarly, we get

$$T(\alpha_1 u_1 + \ldots + \alpha_n u_n) = \alpha_1 T(u_1) + \ldots + \alpha_n T(u_n).$$

Example 7: Check whether $T : V_3 \to V_3$ defined by $T(1, 2, 2) = (2, 3, 1)$, $T(0, 1, 2) = (1, -1, 3)$ and $T(3, -4, 1) = (1, 1, -2)$ and $T(3, -1, 5) = (4, 3, 2)$ is linear.

We know that 4 vectors in V_3 are L.D. Now to check whether first 3 vectors are linearly independent is to be checked.

Using row reduction method, $\begin{bmatrix} 1 & 2 & 2 \\ 0 & 1 & 2 \\ 2 & 3 & 1 \end{bmatrix} \sim \begin{bmatrix} 1 & 2 & 2 \\ 0 & 1 & 2 \\ 0 & -1 & -3 \end{bmatrix}$

$\begin{bmatrix} 1 & 2 & 2 \\ 0 & 1 & 2 \\ 0 & 0 & 1 \end{bmatrix} \sim \begin{bmatrix} 1 & 0 & -2 \\ 0 & 1 & 2 \\ 0 & 0 & 1 \end{bmatrix} \sim \begin{bmatrix} 1 & 0 & 0 \\ 0 & 1 & 0 \\ 0 & 0 & 1 \end{bmatrix}.$

Row rank is 3, these vectors are L.I., therefore writing 4th vector as a linear combination of other 3 vectors, we get $(3, -1, 5) = (1, 2, 2) + (0, 1, 2) + (3, -4, 1)$.

If T is linear, then

$$T(3, -1, 5) = T(1, 2, 2) + T(0, 1, 2) + T(3, -4, 1).$$

$$= (2, 3, 1) + (1, -1, 3) + (1, 1, -2) = (4, 3, 2).$$

This agrees with the given value of $T(3, -1, 5) = (4, 3, 2)$. Hence T is linear.

Example 8: Check whether the map $T : V_2 \to V_2$ defined by $T(1, 2) = (2, 3)$, $T(0, 1) = (1, -1)$ and $T(3, -4) = (5, 7)$ is linear.

V_2 is of dimension 2, therefore two vectors $(1, 2)$, $(0, 1)$, which are linearly independent form a basis for V_2.

Third vector $(3, -4)$ is linear combination of other two vectors, because $(3, -4) = \alpha(1, 2) + \beta(0, 1) = (\alpha, 2\alpha + \beta)$.

On equating $\alpha = 3$, $2\alpha + \beta = -4 \Rightarrow \alpha = 3, \beta = -10$, i.e., $(3, -4) = 3(1, 2) - 10(0, 1)$.

$T(3, -4) = 3\, T(1, 2) - 10\, T(0, 1)$ if T is linear
$= 3(2, 3) - 10(1, -1) = (6, 9) - (10, -10) = (-4, 19) \neq (5, 7)$ as given

Therefore T is not linear.

4.2 NULL AND RANGE SPACES

Definition 4.2: Let $T : U \to V$ be a linear transformation from a vector space U to a vector space V.

Null set of T or kernel of T is denoted and defined by

$$N(T) = \{u \in U \mid T(u) = 0_V\} = \text{Ker } (T).$$

Similarly Range set of linear transformation T from vector space U to a vector space V is denoted and defined by range of $T = R(T) = \{v \in V \mid T(u) = v \text{ for some } u \in U\}$.

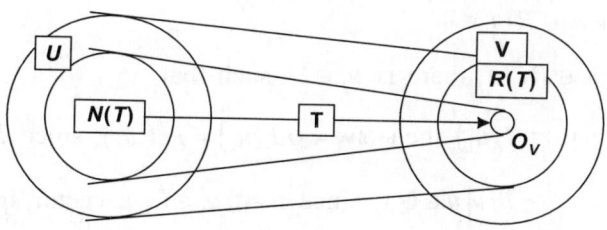

Theorem 4.2: Let $T : U \to V$ be linear transformation from a vector space U to a vector space V.

Then (a) $N(T)$ is a subspace of U.
     (b) $R(T)$ is a subspace of V.

(a) By definition $N(T)$ is a subset of U. So we have to prove only two conditions i.e. $N(T)$ is closed under addition and scalar multiplication.

Let $u_1, u_2 \in N(T)$ i.e. $T(u_1) = 0_V, T(u_2) = 0_V$ by definition of $N(T)$

$$T(u_1 + u_2) = T(u_1) + T(u_2), \quad \text{since } T \text{ is linear}$$
$$= 0_V + 0_V, \quad \text{since } u_1, u_2 \in N(T)$$
$$= 0_V.$$

This shows that $u_1 + u_2 \in N(T)$.

Now for any $u_1 \in N(T)$

$$T(\alpha u_1) = \alpha T(u_1), \quad \text{since } T \text{ is linear}$$
$$= \alpha 0_V, \quad \text{since } u_1 \in N(T)$$
$$= 0_V$$

i.e. $\alpha u_1 \in N(T)$ for all $u_1 \in N(T)$ and any scalar α.

Therefore $N(T)$ is a subspace of U.

(b) To prove $R(T)$ is a subspace of V.

Let $v_1, v_2 \in R(T)$ i.e. there are $u_1, u_2 \in U$ such that $T(u_1) = v_1$ and $T(u_2) = v_2$.

Now $\quad v_1 + v_2 = T(u_1) + T(u_2) = T(u_1 + u_2)$ since T is linear

$$= T(u \mid \text{some } u = u_1 + u_2 \in U), \text{ since } U \text{ is a vector space}$$
$$= T(u) \text{ for some } u \in U.$$

Since $v_1, v_2 \in R(T)$, there is some $u \in U$ such that $T(u) = v_1 + v_2$.

Therefore $v_1 + v_2 \in R(T)$

Now for any $v_1 \in R(T)$, there is $u_1 \in U$ such that

$$v_1 = T(u_1) \text{ then } \alpha v_1 = \alpha T(u_1) = T(\alpha u_1), \text{ since } T \text{ is linear}$$
$$= T(u \mid u \in U), \text{ since } u = \alpha u_1 \in U \text{ a vector space}$$
$$= T(u) \text{ for some } u \in U$$

Therefore $\alpha v_1 \in R(T)$.

Therefore $R(T)$ is closed under addition of vectors and scalar multiplication.

Hence $R(T)$ is a subspace of V.

Now, onward, $N(T)$ and $R(T)$ are used as subspaces.

Dimension of $N(T) = n(T)$ is called nullity of T.
Dimension of $R(T) = r(T)$ is called rank of T.

As an application of null space of a linear map, consider system of m linear equations $A\mathbf{x} = \mathbf{b}$ in n-unknowns, where $A = (a_{ij})_{m \times n}$, $x = (x_1, x_2, ..., x_n)^T$ and $b = (b_1, b_2, ..., b_n)^T$.

Solving this system of linear equations for $b = 0$ is equivalent to finding null space of A, since A maps vector x to 0_v vector.

If $m = n$ and $N(T) = \{0_U\}$, then solution of $Ax = 0_v$ is trivial, but if $N\{T\} \neq \{0_U\}$, we get a non-trivial solution.

Example 9: Let a linear map $T : P_3 \to P_2$ be defined by $T(P)(x) = p'(x)$. Find $N(T)$.

Consider
$$p(x) = a_0 + a_1 x + a_2 x^2 + a_3 x^3 \in P_3$$
$$TP(x) = p'(x) = a_1 + 2a_2 x + 3a_2 x^2$$
$$N(T) = \{p(x) \in P_3(x) | p'(x) = 0\}$$
$$= \{p(x) \in P_3(x) | a_1 + 2a_2 x + 3a_3 x^2 = 0 \text{ for all } x\}$$
$$= (p(x) | a_1 + 2a_2 x + 3a_3 x^2 = 0 \Rightarrow a_1 = a_2 = a_3 = 0)$$
$$= \{a_0\}, \text{ since } a_1 = a_2 = a_3 = 0$$
$$= [1], \text{ for any real number } a_0$$
$$= \text{set of polynomials of degree zero including 0.}$$

Therefore $n(T) = 1$. $R(T) = P_2$ Therefore $r(T) = 3$.

Example 10: Let $T : V_2 \to V_3$ be a linear map defined by $T(x_1, x_2) = (x_1 - x_2, 0, 0)$. To find null space and range space of T.

$$N(T) = \{(x_1, x_2) | T(x_1, x_2) = (0, 0, 0)\}$$
$$= \{(x_1, x_2) | x_1 - x_2 = 0\}$$
$$= \{(x_1, x_1) | x_1 \text{ is any real number}\}$$
$$= \{x_1 (1, 1), x_1 \text{ any real number}\} = [(1, 1)].$$

Therefore $n(T) = 1$.
$$R(T) = \{(x_1 - x_2, 0, 0) | x_1 \text{ and } x_2 \text{ are any real nos.}\}$$
$$= \{(x_1 - x_2) (1, 0, 0)\}$$
$$= [(1, 0, 0)], \text{ Therefore } r(T) = 1.$$

Example 11: Let A linear map $T : V_3 \to V_2$ be defined by
$T(x_1, x_2, x_3) = (x_1 - 2x_2, x_2 + x_3)$ for all $(x_1, x_2, x_3) \in V_3$. To find the null space and range space of T.

$$N(T) = \{(x_1, x_2, x_3,) \mid T(x_1, x_2, x_3) = (0, 0)\}$$
$$= \{(x_1, x_2, x_3,) \mid (x_1 - 2x_2, x_2 + x_3) = (0, 0)\}$$
$$= \{(x_1, x_2, x_3,) \mid (x_1 - 2x_2 = 0, x_2 + x_3 = 0)\}$$
$$= \{(x_1, x_2, x_3,) \mid (x_1 = 2x_2, x_3 = -x_2)\}$$
$$= \{(2x_2, x_2, -x_2) \mid x_2 \text{ any real number}\}$$
$$= \{x_2(2, 1, -1)\} = [(2, 1, -1)].$$

Therefore Basis of $N(T) = \{(2, 1, -1)\}$. Hence $n(T) = 1$

$$R(T) = \{(x_1 - 2x_2, x_2 + x_3) \mid x_1, x_2, x_3 \text{ any real numbers}\}$$
$$= \{(x_1, 0) + (-2x_2, x_2) + (0, x_3)\}$$
$$= \{x_1(1, 0) + x_2(-2, 0) + x_3(0, 1)\}$$
$$= [(1, 0), (-2, 0), (0, 1)]$$
$$= [(1, 0), (0, 1)], \text{ since } (-2, 0) \text{ is a linear combination of } (1, 0) \text{ and } (0, 1)$$

Therefore basis of $R(T)$ is $= \{(1, 0), (0, 1)\}$ and $r(T) = 2$.

Example 12: Let $T : V_4 \to P_2$ be a linear map from V_4 to P_2, defined by

$$T(x_1, x_2, x_3, x_4) = x_1 + (x_2 + x_3)x + (x_2 - x_4)x^2.$$

Find $N(T)$, $R(T)$, $n(T)$ and $r(T)$ of map T.

$$N(T) = \{(x_1, x_2, x_3, x_4) \mid x_1 + (x_2 + x_3)x + (x_2 - x_4)x^2 = 0\}$$
$$= \{(x_1, x_2, x_3, x_4) \mid x_1 = 0 \quad x_2 + x_3 = 0 \text{ and } x_2 - x_4 = 0\}$$
$$= \{(x_1, x_2, x_3, x_4) \mid x_1 = 0, \quad x_3 = -x_2, \quad x_4 = x_2\}$$
$$= \{(0, x_2, -x_2, x_2) \mid x_2 \text{ any real number}\}$$
$$= [(0, 1, -1, 1)].$$

Therefore basis of $N(T) = \{(0, 1, -1, 1)\}$, and Nullity of $T = \dim N(T) = n(T) = 1$.

$$R(T) = \{x_1 + (x_2 + x_3)x + (x_2 - x_4)x^2 \; \forall \; x_1, x_2, x_3, x_n \in R\}$$
$$= \{x_1 + x_2(x + x^2) + x_3 x - x_4 x^2\}$$
$$= [(1, x + x^2, x, x^2)], \text{ since } x_1, x_2, x_3, x_4 \text{ are any real numbers.}$$

$= [(1, x, x^2)]$, since $x + x^2$ is sum of x and x^2.

$= P_2$.

Therefore $R(T) = P_2$, and rank $(T) = r(T) = 3$.

Definition 4.3:

(a) *A linear map* $T : U \to V$ *is called one-one, if*
$$u_1 \neq u_2 \Rightarrow T(u_1) \neq T(u_2) \ \forall \ u_1, u_2 \in U$$
or equivalently $T(u_1) = T(u_2) \Rightarrow u_1 = u_2$.

(b) A linear map $T : U \to V$ is called onto if for each $v \in V$, there is some $u \in U$ such that $v = T(u)$.

Theorem 4.3: Let $T : U \to V$ be a linear map.

(a) T is one-one iff $N(T) = \{0_U\}$.

(b) If $[u_1, u_2, ..., u_n] = U$, then $R(T) = [T(u_1), T(u_1), ..., T(u_1)]$.

(c) If T is one-one, then $S = \{u_1, u_2, ..., u_n\}$ a subset of U is LI if and only if
$$\{T(u_1), T(u_2), ..., T(u_n)\} \text{ is LI.}$$

Proof:

(a) T is one-one then $N(T) = \{0_U\}$.

Let T is one-one. Now suppose $N(T)$ has a non-zero vector u (say), then by definition of $N(T)$, $T(u) = 0_V$, also $T(0_U) = 0_V$, i.e., $T(u) = 0_V = T(0_U)$, since T is one-one $u = 0_U$. Therefore u can not be a non-zero vector and $N(T) = \{0_U\}$.

Now suppose $N(T) = \{0_U\}$. $T(u_1) = T(u_2) \Rightarrow T(u_1) - T(u_2) = 0_V$, $T(u_1 - u_2) = 0_V$, since T is linear, $u_1 - u_2 = 0_U$ since $N(T) = \{0_U\}$. Therefore $u_1 = u_2$.

Hence $T(u_1) = T(u_2) \Rightarrow u_1 = u_2$, T one-one.

(b) If $[u_1, u_2, ..., u_n] = U$, then $R(T) = [T(u_1), T(u_1), ..., T(u_1)]$.

Let $u \in U$. $u = a_1 u_1 + a_2 u_2 + ... + a_n u_n$.

$T(u) = T(a_1 u_1 + a_2 u_2 + ... + a_n u_n)$

$= a_1 T(u_1) + a_2 T(u_2) + ... + a_n T(u_n)$, since T is a linear map.

$\alpha_1 T(u_1) + \alpha_2 T(u_2) + ... + \alpha_n T(u_n) \in V$. Hence $R(T) = [T(u_1), T(u_1), ..., T(u_1)]$

(c) If T is one-one, then $S = \{u_1, u_2, ..., u_n\}$ a subset of U is LI if and only if
$$\{T(u_1), T(u_2), ..., T(u_n)\} \text{ is LI.}$$

Consider $\alpha_1 T(u_1) + \alpha_2 T(u_2) +, ..., + \alpha_n T(u_n) = 0_V$

$T(\alpha_1 u_1 +, ..., + \alpha_n u_n) = 0_V$, since T is linear, also $T(0_u) = 0_v$,

$\alpha_1 u_1 +, ... + \alpha_n u_n = 0_U$, since T is one-one.

Now if $\{u_1, ..., u_2\}$ is L.I. Then $\alpha_1, \alpha_2, ..., \alpha_n = 0$.

Therefore $\{T(u_1), T(u_2), ..., T(u_n)\}$ is also LI.

Conversely, Suppose $\{T(u_1), T(u_2), ..., T(u_n)\}$ is LI.

Consider $\alpha_1 u_1 + \alpha_2 u_2 +, ..., + \alpha_n u_n = 0_u$.

Now $T(\alpha_1 u_1 + \alpha_2 u_2 +, ..., + \alpha_{n-1} u_{n-1} + \alpha_n u_u) = T(0_u) = 0_v$

$\alpha_1 T(u_1) + \alpha_2 T(u_2) +, ..., + \alpha_{n-1} T(u_{n-1}) + \alpha_n T(u_u) = 0_v$, since T is linear.

Since $\{T(u_1), T(u_2), ..., T(u_n)\}$ is LI, $\alpha_1 = \alpha_2 =, ..., = \alpha_n = 0$.

Therefore set $\{u_1, u_2, ..., u_n\}$ is also LI.

Example 13: Let $T : V_3 \to V_3$ be a linear map defined by
$$T(x_1, x_2, x_3) = (x_1 - x_2, 2x_2 + x_3, 0).$$

Find $N(T)$, $R(T)$ and check whether T is one-one and onto.

To find $\quad N(T) = \{(x_1, x_2, x_3) \in V_3 | T(x, x_2, x_3) = (0, 0, 0)\}$.
$\quad\quad\quad\quad = \{(x_1, x_2, x_3) \in V_3 \div (x_1 - x_2, 2x_2 + x_3, 0) = (0, 0, 0)$.

On equating $x_1 - x_2 = 0$, $2x_2 + x_3 = 0 \Rightarrow x_1 = x_2$, $2x_2 = -x_3$,
$\quad\quad\quad\quad = \{(x_2, x_2, -2x_2) | x_2 \text{ is any real number}\}$
$\quad\quad\quad\quad = [(1, 1, -2)]$

Therefore Basis of $N(T)$ is $\{(1, 1, -2)$ and dim $N(T) = 1$. Linear transformation is not one-one.

$R(T) = \{(x_1 - x_2, 2x_2 + x_3, 0) | \text{ for any real numbers } x_1, x_2, x_3\}$
$\quad\quad = \{(x_1, 0, 0) + (-x_2, 2x_2, 0) + (0, x_3, 0)\}$
$\quad\quad = [(1, 0, 0), (-1, 2, 0), (0, 1, 0)]$
$\quad\quad = [(1, 0, 0), (0, 1, 0)]$, since $(-1, 2, 0)$ is linear combination of other two vectors.

Basis of $R(T)$ is $\{(1, 0, 0), (0, 1, 0)\}$, hence $r(T) = 2$.
Therefore $r(T) = 2. \neq 3 = $ dim V_3, and T is not onto map.

Example 14: Let $T : V_3 \to P_2$ be a linear map defined by
$$T(\alpha_1, \alpha_2, \alpha_3) = \alpha_1 + \alpha_3 + (\alpha_2 - \alpha_1)x + (\alpha_2 + \alpha_3)x^2.$$
Find $N(T)$, $R(T)$, $n(T)$ and $r(T)$. Hence check whether T is one-one and onto.

To find $N(T)$, consider $T(\alpha_1, \alpha_2, \alpha_3) = \alpha_1 + \alpha_3 + (\alpha_2 - \alpha_1)x + (\alpha_2 + \alpha_3)x^2 = 0$, for all real values of x.

Equating coefficients $\alpha_1 + \alpha_3 = 0$, $\alpha_2 - \alpha_1 = 0$, $\alpha_2 + \alpha_3 = 0$, and then on solving, the values obtained are $\alpha_2 = \alpha_1 = -\alpha_3$.

$$\begin{aligned} N(T) &= \{(\alpha_1, \alpha_2, \alpha_3) |\ \alpha_2 = \alpha_1 = -\alpha_3\} \\ &= \{(\alpha_1, \alpha_1, -\alpha_1) \text{ for all real numbers } \alpha_1\} \\ &= \{(\alpha_1(1, 1, -1)\} = [(1, 1, -1)]. \end{aligned}$$

Basis of $N(T)$ is $\{(1, 1, -1)\}$. Hence $n(T) = 1$.

Hence T is not one-one, because $N(T) \neq 0$.

$$\begin{aligned} R(T) &= \{\alpha_1 + \alpha_3 + (\alpha_2 - \alpha_1)x + (\alpha_2 + \alpha_3)x^2 \mid \text{ for real numbers } \alpha_1, \alpha_2, \alpha_3\} \\ &= \{\alpha_1(1 - x) + \alpha_2(x + x^2) + \alpha_3(1 + x^2)\} \\ &= [1 - x, x + x^2, 1 + x^2] \\ &= [1 - x, x + x^2], \text{ since } 1 + x^2 \text{ is sum of first two.} \end{aligned}$$

Basis of $R(T)$ is $\{1 - x, x + x^2\}$, $r(T) = 2$.

$r(T) = 2. \neq 3 = \dim P_2$, hence T is not onto map.

Theorem 4.4: (Rank-Nullity Theorem) Let $T : U \to V$ be a linear map from a finite dimensional vector U to a vector space V. Then $\dim R(T) + \dim N(T) = \dim U$ i.e.
$$r(T) + n(T) = \dim U.$$

Let $\dim U = n$ and $n(T) = r \leq n$. Suppose $B = \{u_1, u_2, ..., u_r\}$, be a basis of $N(T)$.

Extend the basis $B = \{u_1, u_2, ..., u_r\}$ by introducing vectors $u_{r+1}, u_{r+1}, u_{r+2}, ..., u_n$, to the basis

$$B_1 = \{u_1, u_2, ..., u_r, u_{r+1}, ..., u_n\} \text{ of } U, \text{ since } \dim V = n.$$

For $u \in U$, $u = \alpha_1 u_1 + \alpha_2 u_2 + , ..., + \alpha_n u_n$ for some scalars $\alpha_1, \alpha_2, ..., \alpha_n$.

Now $T(u) = T(\alpha_1 u_1 + \alpha_2 u_2 + , ..., + \alpha_n u_n)$

$$= \alpha_1 T(u_1) + \alpha_2 T(u_2) + , ..., \alpha_r T(u_r) + \alpha_{r+1} T(u_{r+1}) + , ..., + \alpha_n T(u_n).$$

Since B is basis of $N(T)$, $T(u_1) = T(u_2) = ... = T(u_r) = 0_V$, therefore $T(u) = \alpha_1 0_V + \alpha_2 0_V = , ..., + \alpha_r 0_V + \alpha_{r+1} T(u_{r+1}) + , ..., + \alpha_n T(u_n)$

$$= \alpha_{r+1} T(u_{r+1}) + \alpha_{r+2} T(u_{r+2}) + , ..., + \alpha_n T(u_n).$$

Since $T(u) \in R(T)$, $\{T(u_{r+1}), T(u_{r+2}), ..., T(u_n)\}$ spans $R(T)$.

Let $B_2 = \{T(u_{r+1}), T(u_{r+2}), ..., T(u_n)\}$.

Therefore $[T(u_{r+1}), T(u_{r+2}), ..., T(u_n)] = R(T)$.

Now to prove that B_2 is LI. Consider

$$\beta_{r+1} T(u_{r+1}) + \beta_{r+2} T(u_{r+2}) + , ..., + \beta_n T(u_n) = 0_V, \text{ since } T \text{ is linear}$$

$$T(\beta_{r+1} u_{r+1} + \beta_{r+2} u_{r+2} + , ..., + \beta_n u_n) = 0_V$$

$\Rightarrow \quad \beta_{r+1} u_{r+1} + \beta_{r+2} u_{r+2} + , ..., + \beta_n u_n = 0_V$

i.e. $\quad \beta_{r+1} u_{r+1} + , ..., + \beta_n u_n \in N(T)$.

Further $0u_1 + 0u_2 + , ..., + 0u_r + \beta_{r+1} u_{2+1} + , ..., + \beta_n u_n = 0_U$.

And $\{u_1, u_2, ..., u_r, u_n\}$ is LI, therefore all scalars $\beta_{r+1} = \beta_{r+2} = , ..., = \beta_n = 0$.

Hence $\{T(u_{r+1}), ..., T(u_n)\}$ is LI.

This shows that dim $R(T) = n - r$, i.e. $r(T) = n - r$

$$n(T) + r(T) = r + n - r = n = \dim U. \text{ Proved.}$$

Note: Let the Linear system of equation be $A\mathbf{x} = \mathbf{b}$, where A is $n \times n$ coefficient matrix, if $\mathbf{b} = 0$, then solution of $A\mathbf{x} = 0$ is null space of A.

Therefore non-trivial solution of $A\mathbf{x} = 0$ is as finding the null space A.

A is a linear map, would be proved linear later.

Example 15: Let $T : V_3 \to V_4$ be a linear map, defined by

$$T(x_1, x_2, x_3) = (x_1, x_1 + x_2, x_1 + x_2 + x_3, x_3).$$

Find $N(T)$, $R(T)$, $n(T)$ and $r(T)$.

$$N(T) = \{(x_1, x_2, x_3) | x_1 = 0, x_1 + x_2 = 0, x_1 + x_2 + x_3 = 0, x_3 = 0\}$$

$$= \{(0, 0, 0)\} = \{0_{V_3}\}.$$

Therefore $n(T) = 0$

Now $R(T) = \{(x_1, x_1 + x_2, x_1 + x_2 + x_3, x_3) |\ x_1, x_2, x_3$ any real numbers$\}$

$= \{(x_1, x_1, x_1, 0) + (0, x_2, x_2, 0) + (0, 0, x_3, x_3)\}$

$= \{x_1(1, 1, 1, 0) + x_2(0, 1, 1, 0) + x_3(0, 0, 1, 1) | x_1, x_2, x_3$

any real numbers$\}$

$= [(1, 1, 1, 0), (0, 1, 1, 0), (0, 0, 1, 1)].$

Span of three $\{(1, 1, 1, 0), (0, 1, 1, 0), (0, 0, 1, 1)\}$ linearly independent vectors.
Therefore $r(T) = 3$. Now to check $n(T) + r(T) = \dim U = \dim V_3$.
We note that $0 + 3 = 3$, which is true.

Example 16: Let $T : P_2 \to V_2$ be a linear transformation from space P_2 to V_2 defined by $T(\alpha + \beta x + \gamma x^2) = (\alpha - \beta, \beta - \gamma)$.

Find $N(T)$, $R(T)$, $n(T)$ and $r(T)$.

$N(T) = \{\alpha + \beta x + \gamma x^2 | \alpha - \beta = 0, \beta - \gamma = 0\}$

$= \{\alpha + \beta x + \gamma x^2 | \alpha = \beta = \gamma\}$

$= \{\alpha(1 + x + x^2)\}$

$= [(1 + x + x^2)],$ span of a single vector $(1 + x + x^2)$.

Therefore $\dim N(T) = n(T) = 1$.

$R(T) = \{(\alpha - \beta, \beta - \gamma)\} = \{(\alpha, 0) + (-\beta, \beta) + (0, -\gamma)\}$

$= \{\alpha(1, 0) + \beta(-1, 1) + \gamma(0, -1)\}$

$= [(1, 0), (-1, 1), (0, -1)].$

Three vectors in V_2 are linearly dependent, obviously $(-1, 1)$ is linear combination of $(1, 0)$ and $(0, -1)$.

Therefore $R(T) = [(1, 0), (0, -1)]$, now set $\{(1, 0), (0, -1)\}$ is LI.

Hence dim $R(T) = 2 = r(T)$.

Rank and nullity theorem dim $N(T)$ + dim $R(T)$ = dim U = dim P_2

$$1 + 2 = 3.$$ Rank and nullity theorem verified.

Example 17: Let $\{e_1, e_2, e_3, e_4\}$ be standard basis for V_4 and $T : V_4 \to V_3$ be a linear map defined by $T(e_1) = (1, 1, 1)$, $T(e_2) = (1, -1, 1)$, $T(e_3) = (1, 0, 0)$, and $T(e_4) = (1, 0, 1)$. Find the null space $N(T)$, range space $R(T)$, nullity $n(T)$ and rank $r(T)$.

To find $N(T)$, $\quad T(x_1, x_2, x_3, x_4) = (0, 0, 0)$

$$\Rightarrow T(x_1 e_1 + x_2 e_2 + x_3 e_3 + x_4 e_4)$$

$$= x_1 T(e_1) + x_2 T(e_2) + x_3 T(e_3) + x_4 T(e_4)$$

$$= x_1 (1, 1, 1) + x_2 (1, -1, 1) + x_3 (1, 0, 0) + x_4 (1, 0, 1)$$

$$= (x_1 + x_2 + x_3 + x_4, x_1 - x_2, x_1 + x_2 + x_4)$$

$$= (0, 0, 0), \text{ for null space}$$

i.e. $\quad x_1 + x_2 + x_3 + x_4 = 0, \quad x_1 - x_2 = 0, \quad x_1 + x_2 + x_4 = 0$

From first and last relations $x_3 = 0$, from second

$x_1 = x_2, \quad x_4 = -x_1 - x_1 \Rightarrow x_2 = x_1, \quad x_3 = 0, \quad x_4 = -2x_1$.

Therefore $N(T) = \{(x_1, x_1, 0, -2x_1) | x_1 \text{ any real number}\}$

$$= \{x_1(1, 1, -0, 2)\} \text{ for any real number } x_1$$

Therefore $N(T) = [(1, 1, -0, 2)], \quad n(T) = 1$.

Now $\quad R(T) = \{(x_1 + x_2 + x_3 + x_4, x_1 - x_2, x_1 + x_2 + x_3)\}$

$$= [(1, 1, 1), (1, -1, 1), (1, 0, 0), (1, 0, 1)], \text{ in } V_3 \text{ any 4 vectors are LD}.$$

We note that $(1, 0, 1) = \dfrac{1}{2}\{(1, 1, 1) + (1, -1, 1)\}$

Hence removing last vector, $R(T) = [(1,1,1), (1,-1,1), (1,0,0)]$.

To check linear independence of $\{(1,1,1), (1,-1,1), (1,0,0)\}$,
consider $\alpha(1,1,1) + \beta(1,-1,1) + \gamma(1,0,0) = (0,0,0)$

On equating $\alpha + \beta + \gamma = 0$, $\alpha - \beta = 0$, $\alpha + \beta = 0$.

Therefore $\alpha = 0 = \beta = \gamma$.

Hence above set is linearly independent. $R(T) = V_3$, $r(T) = 3$.

From Rank and nullity theorem also
$$n(T) + r(T) = \dim V_4$$
$$1 + r(T) = 4 \Rightarrow r(T) = 3.$$

Therefore Rank $T = 3$.

Example 18: Let a linear map $T : V_3 \to V_3$ be defined by
$$T(x, y, z) = (x + 2y - z, y + z, x + y - 2z).$$

Find a basis and dimension of the null space of the given linear map, also find a basis and dimension of $R(T)$.

For null space $T(z, y, z) = (0, 0, 0) \Rightarrow x + 2y - z = 0$, $y + z = 0$, $x + y - 2z = 0$.
On solving we get $x = 3z$, $y = -z$.

Therefore $N(T) = \{(x, y, z)\} = \{(3z, -z, z)\} = [(3, -1, 1)]$.

Therefore basis of $N(T) = \{(3, -1, 1)\}$ and $\dim N(T) = n(T) = 1$.

Further $R(T) = \{(x + 2y - z, y + z, x + y - 2z) | x, y, z \in R\}$
$$= \{x(1, 0, 1) + y(2, 1, 1) + z(-1, 1, -2) | x, y, z \in R\}$$
$$= [(1, 0, 1), (2, 1, 1), (-1, 1, -2)].$$

We note that $(1, 0, 1) = \frac{1}{3}(2, 1, 1) - \frac{1}{3}(-1, 1, -2)$ i.e. $(1, 0, 1)$ is linear combination of other two vectors.

Therefore these three vectors are LD. Removing $(1, 0, 1)$, from this set, we get
$$R(T) = [(2, 1, 1), (-1, 1, -2)].$$

Hence basis of $R(T) = \{(2, 1, 1), (-1, 1, -2)\}$ and $r(T) = 2$.

4.3 INVERSE LINEAR TRANSFORMATIONS

We have studied linear transformations from a vector space U to a vector space V. Similarly we can study some linear maps from the same vector space V to the same vector space U, without having any references to the linear transformation from U to V. Now suppose given that $T : U \rightarrow V$ is a linear map. We want to know whether a map $S : V \rightarrow U$ exists, which has some relation with T, particularly given $T(u) = v$, can we have a linear map S such that $S(v) = u$ for all $v \in V$. In this section such transformations are to be studied.

Definition 3.4: Let $T : U \rightarrow V$ be a linear transformation, if T is one-one and onto, then T is called non-singular or isomorphic transformation.

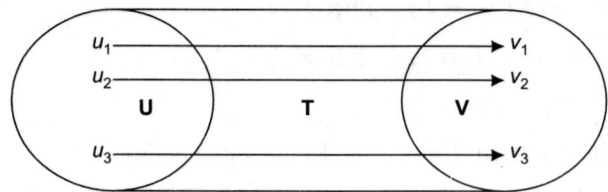

Theorem 4.5: Given a non-singular linear map $T : U \rightarrow V$, then a map $S : V \rightarrow U$ defined by $S(v) = u$ exists and is linear, which is called inverse map of T, whenever $T(u) = v$ for all $u \rightarrow U$ and corresponding $v \in V$.

S is also non-singular and such a map S is denoted by T^{-1} i.e. $S \equiv T^{-1}$.

Now T^{-1} is being used for S in the following proof.

To prove T^{-1} is also linear non-singular and $\left(T^{-1}\right)^{-1} = T$.

Let $u_1, u_2 \rightarrow U$ and $T(u_1) = v_1, T(u_2) = v_2$ where $v_1, v_2 \in V$, then by definition $v_1 = T^{-1}(u_1), v_2 = T^{-1}(u_2)$.

Now $T^{-1}(\alpha v_1 + \beta v_2) = T^{-1}(\alpha T(u_1) + \beta T(u_2))$

$\qquad = T^{-1}(T(\alpha u_1) + T(\beta u_2))$, since T is linear

$\qquad = T^{-1}T(\alpha u_1 + \beta u_2)$ by definition of inverse of T.

$\qquad = \alpha u_1 + \beta u_2$, since $T^{-1}T = I_U$

$\qquad = \alpha T^{-1}(v_1) + \beta T^{-1}(v_2)$.

Therefore T^{-1} is linear.

For each $v \in V$, there is some $u \in U$ such that $T(u) = v$, since T is onto.

Further T is one-one, $u = T^{-1}v$ by definition of T^{-1}. This shows that each $u \in U$ is image of some $v \in V$ under T^{-1}, therefore T^{-1} is onto.

Let $T^{-1}(v_1) = T^{-1}(v_2)$ for some $v_1, v_2 \in V \Rightarrow u_1 = u_2$ by definition.

Applying T on both, we get $T(u_1) = T(u_2)$

i.e., $v_1 = v_2$. Hence T^{-1} is one-one.

To show $(T^{-1})^{-1} = T$, $Tu = v$, i.e., $u = T^{-1}v = T^{-1}T(u)$ for each $u \in U$.

i.e. $T^{-1}T = I_U \Rightarrow (T^{-1})^{-1} = T$.

Theorem 4.6: Let $T : U \to V$ be a linear map and dim U = dim $V = n$. T is one-one if and only if T is onto. (Note U and V are finite-dimensional).

Proof: From $n(T) + r(T) =$ dim U, if T is onto, then $R(T) = V$, so $r(T) = n \Rightarrow n(T) = 0 \Rightarrow T$ is one-one.

If T is one-one $\Rightarrow n(T) = 0$. Now from above result
$R(T) =$ dim (U), but dim U = dim $V = n$,
therefore $R(T) = V$, hence T is onto.

Remark: If $n(T) = 0$, and T is one-one, then T may not be onto in general. Consider the linear map $T(x_1, x_2, ..., x_n) = (0, x_1, x_2 ...)$.

This map is one-one but not onto.

Example 19: Prove that the linear map $T : V_3 \to V_3$ defined by $T(e_1) = e_1 + e_2$, $T(e_2) = e_2 + e_3$, and $T(e_3) = e_1 + e_2 + e_3$ is nonsingular, hence find its inverse.

For $u = (x_1, x_2, x_3) \in V_3$, $Tu = T(x_1, x_2, x_3) = T(x_1 e_1 + x_2 e_2 + x_3 e_3)$
$= (x_1 Te_1 + x_2 Te_2 + x_3 Te_3) = (x_1 + x_3, x_1 + x_2 + x_3, x_2 + x_3)$

To check transformation T is one-one, let $T(x_1, x_2, x_3) = (0, 0, 0) \Rightarrow$
$(x_1 + x_3, x_1 + x_2 + x_3, x_2 + x_3) = (0, 0, 0)$.

On equating we get $x_1 + x_3 = 0$, $x_1 + x_2 + x_3 = 0$, and $x_2 + x_3 = 0$.
Solving these equations, we get $x_1 = 0 = x_2 = x_3 = 0$.

Therefore $N(T) = \{0_{v_3}\}$ and hence T is one-one.

T is also onto since dimension of both spaces is same.

Hence, T is nonsingular and T^{-1} exist

Now two methods are given to find T^{-1}, the inverse map of T, T^{-1} is also a linear, one-one, and onto map from V_3 to V_3.

Method 1: We have $T(e_1) = e_1 + e_2$, $T(e_2) = e_2 + e_3$, $T(e_3) = e_1 + e_2 + e_3$.

Taking inverse on both sides and using T^{-1} linear, we get

$$e_1 = T^{-1}(e_1 + e_2) = T^{-1}(e_1) + T^{-1}(e_2),$$
$$e_2 = T^{-1}(e_2 + e_3) = T^{-1}(e_2) + T^{-1}(e_3),$$
$$e_3 = T^{-1}(e_1 + e_2 + e_3) = T^{-1}(e_1) + T^{-1}(e_2) + T^{-1}(e_3)$$

Solving these three equations for $T^{-1}(e_1)$, $T^{-1}(e_2)$, and $T^{-1}(e_3)$, we get $T^{-1}(e_1) = e_3 - e_2 = (0, -1, 1)$, $T^{-1}(e_2) = e_1 + e_2 - e_3 = (1, 1, -1)$ and $T^{-1}(e_3) = e_3 - e_1 = (-1, 0, 1)$.

Now T^{-1} is constructed linearly as:

$$T^{-1}(x_1, x_2, x_3) = T^{-1}(x_1 e_1 + x_2 e_2 + x_3 e_3)$$
$$= x_1 T^{-1}(e_1) + x_2 T^{-1}(e_2) + x_3 T^{-1}(e_3)$$
$$= x_1(e_3 - e_2) + x_2(e_1 + e_2 - e_3) + x_3(e_3 - e_1)$$
$$= x_1(0, -1, 1) + x_2(1, 1, -1) + x_3(-1, 0, 1)$$
$$= (x_2 - x_3, x_2 - x_1, x_1 - x_2 + x_3).$$

Hence $T^{-1}(x_1, x_2, x_3) = (x_2 - x_3, x_2 - x_1, x_1 - x_2 + x_3)$.

Method 2: Given that $T(x_1, x_2, x_3) = (x_1 + x_3, x_1 + x_2 + x_3, x_2 + x_3)$,

Suppose $T(x_1, x_2, x_3) = (y_1, y_2, y_3)$. i.e., $T^{-1}(y_1, y_2, y_3) = (x_1, x_2, x_3)$.

Now our aim is to find x_1, x_2, x_3 in terms of y_1, y_2, y_3.

Therefore, on equating the values of $T(x_1, x_2, x_3)$ from two different forms, we get $\quad x_1 + x_3 = y_1, \ x_1 + x_2 + x_3 = y_2, \ x_2 + x_3 = y_3$.

On solving these equations for x_1, x_2, x_3, in terms of y_1, y_2, y_3, we get $\quad x_1 = y_2 - y_3, \ x_2 = y_2 - y_1, \ x_3 = y_1 + y_3 - y_2$.

$$T^{-1}(y_1, y_2, y_3) = (x_1, x_2, x_3) = (y_2 - y_3, y_2 - y_1, y_1 + y_3 - y_2).$$

Hence $T^{-1}(x_1, x_2, x_3) = (x_2 - x_3, x_2 - x_1, x_1 - x_2 + x_3)$.

Example 20: Let $T : P_2 \to P_2$ be defined by

$$T(a_0 + a_1 x + a_2 x^2) = (a_1 + a_2) + (a_0 + a_2)x + (a_0 + a_1)x^2$$

Prove that T is one-one and onto, hence find T^{-1}.

To find null space $N(T)$, let $T(a_0 + a_1 x + a_2 x^2) = 0$.

$$(a_1 + a_2) + (a_0 + a_2)x + (a_0 + a_1)x^2 = 0.$$

On equating the coefficients of $1, x, x^2$,

we get $\quad a_1 + a_2 = 0, \ a_0 + a_2 = 0, \ a_0 + a_1 = 0$.

On solving we get $a_0 = 0$, $a_1 = 0$ and $a_2 = 0$.

Therefore $N(T) = \{0\}$.

T is one-one and T is onto also because dimension of both spaces is 3
Hence T^{-1} exists. Now to find T^{-1}.

Method 1: Choosing $a_0 = 1$, $a_1 = a_2 = 0$, $T(1) = x + x^2$, $a_0 = 0$, $a_1 = 1$, $a_2 = 0$, $T(x) = 1 + x^2$ and $a_0 = 0$, $a_1 = 0$, $a_2 = 1$, $T(x^2) = 1 + x$.

Since T^{-1} exists and is linear, We have $1 = T^{-1}x + T^{-1}x^2$, $x = T^{-1}1 + T^{-1}x^2$ and $x^2 = T^{-1}1 + T^{-1}x$.

On solving for $T^{-1}1$, $T^{-1}x$, $T^{-1}x^2$, we get $2T^{-1}(1) = x + x^2 - 1 \Rightarrow$

$$T^{-1}(1) = \frac{1}{2}(x + x^2 - 1), \quad T^{-1}(x) = \frac{1}{2}(1 + x^2 - x) \text{ and } T^{-1}(x^2) = \frac{1}{2}(1 + x - x^2).$$

Therefore $T^{-1}(b_0 + b_1 x + b_2 x^2) = b_0 T^{-1}1 + b_1 T^{-1} x + b_2 T^{-1}x^2$, since T^{-1} is linear.
Hence

$$T^{-1}(b_0 + b_1 x + b_2 x^2) = \frac{b_0}{2}(x + x^2 - 1) + \frac{b_1}{2}(1 + x^2 - x) + \frac{b_2}{2}(1 + x - x^2).$$

$$= \frac{1}{2}(b_1 + b_2 - b_0) + \frac{1}{2}(b_0 + b_2 - b_1)x + \frac{1}{2}(b_0 + b_1 - b_2)x^2.$$

$$T^{-1}(a_0 + a_1 x + a_2 x^2) = \frac{1}{2}(a_1 + a_2 - a_0) + \frac{1}{2}(a_0 + a_2 - a_1)x$$

$$+ \frac{1}{2}(a_0 + a_1 - a_2)x^2$$

Method 2: Let $T(a_0 + a_1 x + a_2 x^2) = b_0 + b_1 x + b_2 x^2$ and given map is
$$T(a_0 + a_1 x + a_2 x^2) = (a_1 + a_2) + (a_0 + a_2)x + (a_0 + a_1)x^2$$
i.e., $\quad b_0 + b_1 x + b_2 x^2 = (a_1 + a_2) + (a_0 + a_2)x + (a_0 + a_1)x^2$

On equating the coefficients of x on both sides, we get $a_1 + a_2 = b_0$, $a_0 + a_2 = b_1$ and $a_0 + a_1 = b_2$,

On solving we get $a_0 = \frac{1}{2}(b_1 + b_2 - b_0)$, $a_1 = \frac{1}{2}(b_0 + b_2 - b_1)$ and $a_2 = \frac{1}{2}(b_0 + b_1 - b_2)$.

$$T^{-1}(b_0 + b_1 x + b_2 x^2) = \frac{1}{2}(b_1 + b_2 - b_0) + \frac{1}{2}(b_0 + b_2 - b_1)x + \frac{1}{2}(b_0 + b_1 - b_2)x^2.$$

Therefore,

$$T^{-1}(a_0 + a_1 x + a_2 x^2) = \frac{1}{2}(a_1 + a_2 - a_0) + \frac{1}{2}(a_0 + a_2 - a_1)x + \frac{1}{2}(a_0 + a_1 - a_2)x^2.$$

Example 21: Let $T : V_3 \to P_2$ be linear map defined by
$$T(\alpha, \beta, \gamma) = (\alpha + \beta) + (\beta + \gamma)x + (\gamma + \alpha)x^2$$

Prove that T is one-one and onto and hence find T^{-1}.

$$N(T) = \{(\alpha, \beta, \gamma) | (\alpha + \beta) + (\beta + \gamma)x + (\gamma + \alpha)x^2 = 0\}$$
$$= \{(\alpha, \beta, \gamma) | \alpha + \beta = \beta + \gamma = \gamma + \alpha = 0\}$$
$$= \{(\alpha, \beta, \gamma) | \alpha = \beta = \gamma = 0\}$$
$$= \{(0, 0, 0)\} \text{ i.e., } N(T) = \{0_{V_3}\}.$$

Therefore $r(T) = 3$, using Rank and nullity theorem and $R(T) = V_3$, since dim $V_3 = 3$, dim $P_2 = 3$, and dim $N(T) = 0$, hence T is non-singular.

Now to find inverse of T^{-1}.

Method 1: Let $T^{-1}(a + bx + cx^2) = (\alpha, \beta, \gamma) \Rightarrow T(\alpha, \beta, \gamma) = a + bx + cx^2$.
Given linear map $T : V_3 \to P_2$ by $T(\alpha, \beta, \gamma) = (\alpha + \beta) + (\beta + \gamma)x + (\gamma + \alpha)x^2$.
Therefore equating right hand side of $T(\alpha, \beta, \gamma)$ from above two relations,
We get $\quad \alpha + \beta = a, \beta + \gamma = b, \gamma + \alpha = c$.

On solving for $\alpha,$ β, and γ we have, $\alpha = \dfrac{1}{2}(a - b + c)$,

$$\beta = \frac{1}{2}(a + b - c) \text{ and } \gamma = \frac{1}{2}(-a + b + c).$$

Therefore

$$T^{-1}(a + bx + cx^2) = \left(\frac{1}{2}(a - b + c), \frac{1}{2}(a + b - c), \frac{1}{2}(-a + b + c)\right).$$

Method 2: We can also find T^{-1} using standard bases as $B_1 = \{e_1, e_2, e_3\}$ where $e_1 = (1, 0, 0), e_2 = (0, 1, 0)$ and $e_3 = (0, 0, 1)$ for V_3 and $B_2 = \{1, x, x^2\}$ for P_2.

For $\alpha = 1, \beta = 0, \gamma = 0, T(1, 0, 0) = 1 + x^2$, i.e., $T^{-1}(1 + x^2) = e_1 \Rightarrow T^{-1}1 + T^{-1}x^2 = e_1$.

For $\alpha = 0, \beta = 1, \gamma = 0, T(0, 1, 0) = 1 + x$, i.e., $T^{-1}(1 + x) = e_2 \Rightarrow T^{-1}1 + T^{-1}x = e_2$.

For $\alpha = 0, \beta = 0, \gamma = 1, T(0, 0, 1) = x + x^2$, i.e., $T^{-1}(x + x^2) = e_3 \Rightarrow T^{-1}x + T^{-1}x^2 = e_3$.

On solving for $T^{-1}1, T^{-1}x$ and $T^{-1}x^2$ we get

$$T^{-1}1 = \frac{1}{2}(e_1 + e_2 - e_3), \ T^{-1}x = \frac{1}{2}(-e_1 + e_2 + e_3), \ T^{-1}x^2 = \frac{1}{2}(e_1 - e_2 + e_3).$$

Therefore $T^{-1}(a + bx + cx^2) = aT^{-1}1 + bT^{-1}x + cT^{-1}x^2$, since T^{-1} is linear

$$= \frac{a}{2}(e_1 + e_2 + e_3) + \frac{b}{2}(-e_1 + e_2 + e_3) + \frac{c}{2}(e_1 - e_2 + e_3) \text{ on simplification.}$$

$$= \left(\frac{1}{2}(a-b+c), \frac{1}{2}(a+b-c), \frac{1}{2}(-a+b+c)\right).$$

Note: Some important results of linear map are mentioned below for convenience of the readers:

Let $T : U \to V$ be a linear map and dim U = dim $V = n$. Then the following statements are equivalent:

(a) T is nonsingular (an isomorphism)
(b) T is one-one.
(c) T transforms linearly independent subsets of U into linearly independent subsets of V.
(d) T transforms every basis for U into a basis for V.
(e) T is onto.
(f) $r(T) = n$.
(g) $n(T) = 0$.
(h) T^{-1} exists.

4.4 MORE ABOUT LINEAR TRANSFORMATIONS

In this section some more important terms related to linear maps are given briefly for the completeness of the subject.

(a) Addition of Linear maps: Let $T : U \to V$ and $S : U \to V$ be any two linear transformations from a vector space U to a vector space V.

The addition of two linear transformations $T, S : U \to V$ is defined by the transformation $M : U \to V$ as $M(u) = T(u) + S(u)$ for all $u \in U$. M is again a linear map from U to V.

Scalar multiplication of T by a scalar α is defined by $(\alpha T)(u) = \alpha(T(u))$ for all $u \in U$.

Further It can be shown very easily that set of all linear transformations from U to V denoted by $L(U, V)$ is a vector space with vectors as linear transformation with addition of linear transformations and scalar multiplication as defined above and

$$\dim L(U, V) = \dim U \times \dim V.$$

Example 22: Let $T, S : V_3 \to P_2$ be defined by

$T(\alpha, \beta, \gamma) = \alpha + \beta x + \gamma x^2$ and $S(\alpha, \beta, \gamma) = \gamma + \alpha x + \beta x^2$,

then

$(T + S)(\alpha, \beta, \gamma) = T(\alpha, \beta, \gamma) + S(\alpha, \beta, \gamma)$

$\qquad = \alpha + \beta x + \gamma x^2 + \gamma + \alpha x + \beta x^2 = (\alpha + \beta) + (\beta + \gamma)x + (\gamma + \alpha)x^2.$

Scalar multiplication of T by k is $(kT)(\alpha, \beta, \gamma) = k(\alpha + \beta x + \gamma x^2) = k\alpha + k\beta x + k\gamma x^2$.
This can be shown very easily that $(T + S)$ and (kT) are linear maps from V_3 to P_2.

(b) Composition of Linear Maps: Let $T : U \to V$ and $S : V \to W$ be two linear transformations. The composition of S and T, denoted by $(SoT) : U \to W$ is defined by $(SoT)(u) = S(T(u))$ for all $u \in U$. Further it can be proved that composition of two linear transformations is again Linear transformation from $U \to W$. If (SoT) is defined i.e. $SoT : U \to W$, then ToS may not be defined because $(ToS)v = T(Sv) = Tw$ for some $v \in V$ and some $w \in W$, but T is not defined on W. In case if it is defined, then

$SoT \neq ToS$ need not be true in general.

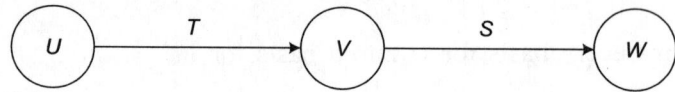

Example 23: Let two linear transformations $T : V_3 \to P_2$ and $S : P_2 \to V_3$ be defined by $T(\alpha, \beta, \gamma) = \alpha + \beta x + \gamma x^2$, and $s(\alpha + \beta x + \gamma x^2) = (\alpha + \beta, \beta + \gamma, \gamma + \alpha)$. Composition of these two maps are:

$(SoT)(\alpha, \beta, \gamma) = S(T(\alpha, \beta, \gamma)) = S(\alpha + \beta x + \gamma x^2) = (\alpha + \beta, \beta + \gamma, \gamma + \alpha)$. This shows that (SoT) is defined from from V_3 to V_3.

$(ToS)((\alpha + \beta x + \gamma x^2) = T(S((\alpha + \beta x + \gamma x^2)) = T(\alpha + \beta, \beta + \gamma, \gamma + \alpha) = \alpha + \beta + (\beta + \gamma)x + (\gamma + \alpha)x^2$. This shows that (ToS) is defined from $P_2 \to P_2$.

Linear transformations (SoT) and (ToS) both are defined but on different spaces, hence can not be equated.

(c) Identity Maps: Let $T : U \to V$ and $S : V \to U$ be non-singular linear transformations with dim U = dim V = n. be defined as

$(SoT)(u) = S(T(u)) = S(v) = u, \forall\ u \in U$, then $SoT = I_U$ identity map on vector space U.

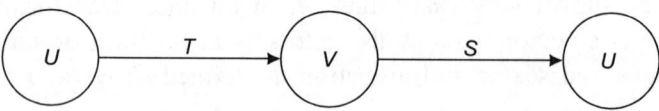

Similarly $(ToS)(v) = T(s(v)) = T(u) = v, \forall\ v \in V$, then $ToS = I_V$, identity map on vector space V.

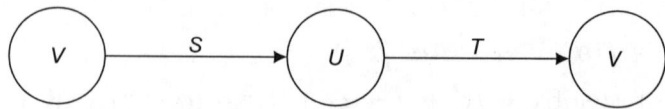

Example 24: Let $T : V_3 \to P_2$ defined by $T(\alpha, \beta, \gamma) = \alpha + \beta x + \gamma x^2$
and $S : P_2 \to V_3$ defined by $S(\alpha + \beta x + \gamma x^2) = (\alpha, \beta, \gamma)$.
$(T o S)((\alpha + \beta x + \gamma x^2) = T(S(\alpha + \beta x + \gamma x^2)) = T(\alpha, \beta, \gamma) = \alpha + \beta x + \gamma x^2$.
This shows $T o S = I_{P_2}$ identity map on P_2.
Similarly $(S o T) (\alpha, \beta, \gamma) = S(T((\alpha, \beta, \gamma)) = S(\alpha + \beta x + \gamma x^2) = (\alpha, \beta, \gamma)$.
This shows that $S o T = I_{V_3}$. Identity map on V_3.

Therefore in both cases identity maps are on different vector spaces.

(d) Idempotent Map: A linear transformation T on a vector V i.e. $T : V \to V$ is said to be idempotent map, if $T^2 u = Tu$ for all $u \in V$ and is written as $T^2 = T$.

Example 25: $T : V_3 \to V_3$ defined by $T(x_1, x_2, x_3) = (x_1, 0, x_3)$

$$T(T(x_1, x_2, x_3) = T(x_1, 0, x_3) = (x_1, 0, x_3))$$

$T^2 = T$. Therefore T is idempotent.

(e) Nilpotent Map: A linear transformation T on a vector space V is said to be nilpotent of order $n > 1$ on V, if $T^n u = 0$ for all $u \in V$ and written as $T^n = 0$.

Example 26: Differential operator D is nilpotent of order $(n + 1)$ on the space P_n of polynomials of degree $\leq n$.

since $D^{n+1} P_n(x) = \dfrac{d^{n+1}}{dx^{n+1}}(P_n(x)) = 0$.

Example 27: Let: $T : V_3 \to V_3$ be defined by

$$T(x_1, x_2, x_3) = (0, x_1, x_2)$$

$$T^2(x_1, x_2, x_3) = T(0, x_1, x_2) = (0, 0, x_1)$$

$$T^3(x_1, x_2, x_3) = T(0, 0, x_1) = (0, 0, 0) = 0_{V_3}.$$

Therefore T is nilpotent linear transformation of order 3.

4.5 MATRICES RELATED TO LINEAR TRANSFORMATION

Let U and V be two finite-dimensional vector spaces of dimensions n and m with ordered bases $B_1 = \{u_1, u_2, ..., u_n\}$ of U and $B_2 = \{v_1, v_2, ..., v_m\}$ of V and further let a linear transformation $T : U \to V$ be defined by

$T(u_j) = \alpha_{1j} v_1 + \alpha_{2j} v_2 + ,..., + \alpha_{mj} v_m, j = 1, 2,....n$, since $T(u_j)$ is a vector in V for u_j in U.

In this $m \times n$ scalars (α_{ij}), $i = 1,...,m$, and $j = 1, 2,...,n$ arranged in the matrix form as

$$\begin{bmatrix} \alpha_{11} & \alpha_{12} & . & . & . & \alpha_{1n} \\ \alpha_{21} & \alpha_{22} & . & . & . & \alpha_{2n} \\ . & & & & & \\ . & & & & & \\ . & & & & & \\ \alpha_{m1} & \alpha_{m2} & . & . & . & \alpha_{mn} \end{bmatrix}$$, with m rows and n columns,

also defines Linear Transformation from U to V completely, because any $u \in U$ is a linear combination of basis vectors of B_1, i.e. $u = \alpha_1 u_1 + \alpha_2 u_2 +,..., + \alpha_n u_n$ for some scalars $\alpha_1, \alpha_2,..., \alpha_n$.

Note: Scalars used in $T(u_j)$ are elements of j column of the matrix.

Now $T(u) = \alpha_1 T(u_1) + \alpha_2 T(u_2) +,..., + \alpha_n T(u_n)$ since T is linear.

$$= [\alpha_1, \alpha_2, ..., \alpha_j ..., \alpha_n] \begin{bmatrix} T(u_1) \\ T(u_2) \\ \\ T(u_j) \\ \\ T(u_n) \end{bmatrix}$$

$$= [\alpha_1, \alpha_2, ..., \alpha_j ..., \alpha_n] \begin{bmatrix} \alpha_{11} v_1 + \alpha_{21} v_2 + ... + \alpha_{i1} v_i + ... + \alpha_{m1} v_m \\ \alpha_{12} v_1 + \alpha_{22} v_2 + ... + \alpha_{i2} v_i + ... + \alpha_{m2} v_m \\ \alpha_{1j} v_1 + \alpha_{2j} v_2 + ... + \alpha_{ij} v_i + ... + \alpha_{mj} v_m \\ \alpha_{1n} v_1 + \alpha_{2n} v_2 + ... + \alpha_{in} v_i + ... + \alpha_{mn} v_m \end{bmatrix}$$

$$= \begin{bmatrix} \alpha_1 \alpha_{11} v_1 + \alpha_1 \alpha_{21} v_2 + ... + \alpha_1 \alpha_{i1} v_i + ... + \alpha_1 \alpha_{m1} v_m \\ \alpha_2 \alpha_{12} v_1 + \alpha_2 \alpha_{22} v_2 + ... + \alpha_2 \alpha_{i2} v_i + ... + \alpha_2 \alpha_{m2} v_m \\ \alpha_j \alpha_{1j} v_1 + \alpha_j \alpha_{2j} v_2 + ... + \alpha_j \alpha_{ij} v_i + ... + \alpha_j \alpha_{mj} v_m \\ \alpha_n \alpha_{1n} v_1 + \alpha_n \alpha_{2n} v_2 + ... + \alpha_n \alpha_{in} v_i + ... + \alpha_n \alpha_{mn} v_m \end{bmatrix}$$

$$= \begin{bmatrix} \alpha_{11}v_1 + \alpha_{21}v_2 + \ldots + \alpha_{i1}v_i + \ldots + \alpha_{m1}v_m \\ \alpha_{12}v_1 + \alpha_{22}v_2 + \ldots + \alpha_{i2}v_i + \ldots + \alpha_{m2}v_m \\ \alpha_{1j}v_1 + \alpha_{2j}v_2 + \ldots + \alpha_{ij}v_i + \ldots + \alpha_{mj}v_m \\ \alpha_{1n}v_1 + \alpha_{2n}v_2 + \ldots + \alpha_{in}v_i + \ldots + \alpha_{mn}v_m \end{bmatrix} \begin{bmatrix} \alpha_1 \\ \alpha_2 \\ \alpha_j \\ \alpha_n \end{bmatrix}$$

$$= [v_1, v_2, \ldots, v_j, \ldots, v_m] \begin{bmatrix} \alpha_{11} & \alpha_{12} & \cdots & \alpha_{1n} \\ \alpha_{21} & \alpha_{22} & \cdots & \alpha_{2n} \\ \vdots & & & \vdots \\ \alpha_{m1} & \alpha_{m2} & \cdots & \alpha_{mn} \end{bmatrix} \begin{bmatrix} \alpha_1 \\ \alpha_2 \\ \alpha_j \\ \alpha_n \end{bmatrix}$$

above is the product of three matrices basis vectors of vector space V, matrix obtained from coefficients of the linear transformation T and coordinate vector of vector u of vector space U.

Definition 4.5: Matrix of linear Transformation:

Therefore scalars used in the above matrix are used to express $T(u_1), (u_2), \ldots, T(u_n)$ as linear combination of basis vectors of V.

Therefore middle $(a_{ij})_{m \times n} = \begin{bmatrix} \alpha_{11} & \alpha_{12} & \cdots & \alpha_{1n} \\ \alpha_{21} & \alpha_{22} & \cdots & \alpha_{2n} \\ \vdots & & & \vdots \\ \alpha_{m1} & \alpha_{m2} & \cdots & \alpha_{mn} \end{bmatrix}$ matrix represents linear

transformation T from U to V, with ordered bases B_1 and B_2 and is denoted by

$$[T : B_1, B_2] = (\alpha_{ij}), \ i = 1, 2 \ldots m \text{ and } j = 1, 2 \ldots n.$$

Example 28: Let $T : P_2 \to V_3$ be a linear map defined by

$$T(a_0 + a_1 x + a_2 x^2) = (a_0, a_2 - a_1, a_2 - a_0),$$

with ordered bases $B_1 = \{1, x, x^2\}$ of P_2 and $B_2 = \{(1, 0, 0), (0, 1, 0), (0, 0, 1)\}$ of V_3.

4.28   *Elementary Linear Algebra*

Write the matrix of linear transformation corresponding to the linear map T.

On taking: $a_0 = 1, a_1 = 0, a_2 = 0$;

$$T(1) = (1, 0, -1) = 1(1, 0, 0) + 0(0, 1, 0) - 1(0, 0, 1)$$

On taking: $a_0 = 0, a_1 = 1, a_2 = 0$;

$$T(x) = (0, -1, 0) = 0(1, 0, 0) - 1(0, 1, 0) + 0(0, 0, 1)$$

On taking: $a_0 = 0, a_1 = 0, a_2 = 1$;

$$T(x^2) = (0, 1, 1) = 0(1, 0, 0) + 1(0, 1, 0) + 1(0, 0, 1).$$

Therefore associated matrix of linear transformation is

$$(T : B_1, B_2) = \begin{bmatrix} 1 & 0 & 0 \\ 0 & -1 & 1 \\ -1 & 0 & 1 \end{bmatrix}.$$

Example 29:   Let $T : V_3 \to P_3$ be a linear map defined by

$$T(\alpha, \beta, \gamma) = \alpha + (\beta + \gamma)x + (\gamma - \alpha)x^2 + \gamma x^3.$$

$B_1 = \{(1, 0, 0), (0, 1, 0), (0, 0, 1)\}$ is an ordered basis of V_3, and
$B_2 = \{1 + x, x + x^2, x^2 + x^3, x^3\}$ is an ordered basis of P_3.

Find the associated matrix of transformation.

On taking: $\alpha = 1, \beta = 0, \gamma = 0$

$$T(1, 0, 0) = 1 - x^2 = 1(1 + x) - 1(x + x^2) + 0(x^2 + x^3) + 0x^3$$

On taking: $\alpha = 0, \beta = 1, \gamma = 0$

$$T(0, 1, 0) = x = 0(1 + x) + 1(x + x^2) - 1(x^2 + x^3) + 1x^3$$

On taking: $\alpha = 0, \beta = 0, \gamma = 1$

$$T(0, 0, 1) = 0 + x + x^2 + x^3$$
$$= 0(1 + x) + 1(x + x^2) + 0(x^2 + x^3) + 1x^3.$$

Therefore, matrix of Transformation is

$$(T : B_1, B_2) = \begin{bmatrix} 1 & 0 & 0 \\ -1 & 1 & 1 \\ 0 & -1 & 0 \\ 0 & 1 & 1 \end{bmatrix}.$$

Now conversely, if we are given ordered bases B_1 and B_2 of two vector spaces U and V respectively and matrix associated with linear map T i.e. $(T, B_1, B_2) = (\alpha_{ij})$, is given then linear map $T : U \to V$ can be determined.

Suppose $B_1 = \{u_1, u_2, ..., u_n\}$ and $B_2 = \{v_1, v_2, ..., v_m\}$ are bases of vector spaces U and V respectively, and corresponding matrix $(T, B_1, B_2) = (\alpha_{ij})$, $i = 1, 2 ... m$, $j = 1, 2 ... n$.)

Now as given

$$T(u_1) = \alpha_{11} v_1 + \alpha_{21} v_2 + ... + \alpha_{m1} v_m$$
$$\vdots \qquad \vdots \qquad \vdots \qquad \vdots$$
$$T(u_j) = \alpha_{1j} v_1 + \alpha_{2j} v_2 + ... + \alpha_{mj} v_m$$
$$\vdots \qquad \vdots \qquad \vdots \qquad \vdots$$
$$T(u_n) = \alpha_{1n} v_1 + \alpha_{2n} v_2 + ... + \alpha_{mn} v_m.$$

And for any $u \in U$, $u = \alpha_1 u_1 + , ..., + \alpha_n u_n$ is expressed a linear combination of vectors of B_1, for some $\alpha_1, \alpha_2, ..., \alpha_{n-1}, \alpha_n$ scalars.

$$T(u) = \alpha_1 T(u_1) + , ..., + \alpha_n T(u_n), \text{ since } T \text{ is linear.}$$

Substituting for $T(u_1), T(u_2), ..., T(u_n)$ from above and simplifying the linear expressions of the vectors, we get

$$T(u) = \alpha_1 (\alpha_{11} v_1 + \alpha_{21} v_2 + ... + \alpha_{m1} v_m) + , ..., + \alpha_n (\alpha_{1n} v_1 + \alpha_{2n} v_2 + ... + \alpha_{mn})$$
$$= (\alpha_1 \alpha_{11} + ... \alpha_n \alpha_{1n}) v_1 + ... + (\alpha_1 \alpha_{m1} + ... \alpha_n \alpha_{mn}) v_m.$$

$$= [v_1, ..., v_m] \begin{bmatrix} \alpha_1 \alpha_{11} + ... \alpha_n \alpha_{1n} \\ \\ \\ \alpha_1 \alpha_{m1} + ... \alpha_n \alpha_{mn} \end{bmatrix}$$

$$= [v_1, ..., v_m] \begin{bmatrix} \alpha_{11} & \alpha_{12} & . & . & . & \alpha_{1n} \\ \alpha_{21} & \alpha_{22} & . & . & . & \alpha_{2n} \\ . & & & & & \\ . & & & & & \\ . & & & & & \\ \alpha_{m1} & \alpha_{m2} & . & . & . & \alpha_{mn} \end{bmatrix} \begin{bmatrix} \alpha_1 \\ \vdots \\ \vdots \\ \vdots \\ \alpha_m \end{bmatrix}.$$

This is same as already discussed.

4.30 Elementary Linear Algebra

Example 30: Let $T : V_3 \to P_2$ be a linear map. Matrix corresponding to the linear map T be

$$(T; B_1, B_2) = \begin{bmatrix} -1 & 2 & 0 \\ 0 & 3 & -2 \\ 1 & -1 & 3 \end{bmatrix},$$

where $B_1 = \{(1,1,0), (0,1,1), (1,0,1)\}$ is a basis of V_3 and
$B_1 = \{1+x, x+x^2, x^2+1\}$ is a basis of P_2.

Find $T(x_1, x_2, x_3)$ for a vector $(x_1, x_2, x_3) \in V_3$.

$$T(1,1,0) = (-1)(1+x) + 0(x+x^2) + 1(x^2+1) = x^2 - x$$

$$T(0,1,1) = 2(1+x) + 3(x+x^2) - 1(x^2+1) = 2x^2 + 5x + 1$$

$$T(1,0,1) = 0(1+x) - 2(x+x^2) + 3(x^2+1) = x^2 - 2x + 3.$$

Now $(x_1, x_2, x_3) = \alpha(1,1,0) + \beta(0,1,1) + \gamma(1,0,1)$
$$= (\alpha + \gamma, \alpha + \beta, \beta + \gamma)$$

on equating the components of the vectors and solving for α, β, γ, we get

$$(x_1, x_2, x_3) = \frac{1}{2}(x_1 + x_2 - x_3)(1,1,0) + \frac{1}{2}(-x_1 + x_2 + x_3)(0,1,1)$$

$$(x_1, x_2, x_3) = \frac{1}{2}(x_1 + x_2 - x_3)(1,1,0) + \frac{1}{2}(-x_1 + x_2 + x_3)(0,1,1)$$

$$+ \frac{1}{2}(x_1 - x_2 + x_3)(1,0,1).$$

Now $T(x_1, x_2, x_3) = \frac{1}{2}(x_1 + x_2 - x_3) T(1,1,0) + \frac{1}{2}(-x_1 + x_2 + x_3) T(0,1,1)$

$$+ \frac{1}{2}(x_1 - x_2 + x_3) T(1,0,1), \text{ since } T \text{ is linear.}$$

$$T(x_1, x_2, x_3) = \frac{1}{2}(x_1 + x_2 - x_3)(x^2 - x) + \frac{1}{2}(-x_1 + x_2 + x_3)(2x^2 + 5x + 1)$$

$$+ \frac{1}{2}(x_1 - x_2 + x_3)(x^2 - 2x + 3)$$

$$= (x_2 + x_3)x^2 + (-4x_1 + 3x_2 + 2x_2)x + (x_1 - x_2 + 2x_3),$$

on simplification.

4.6 RANK OF MATRIX

We have studied that a linear transformation T from an n-dimensional vector space to an m-dimensional vector space can be identified by a $m \times n$ matrix A.

$$A = \begin{bmatrix} a_{11} & a_{12} & \cdots & a_{1j} & \cdots & a_{1n} \\ a_{21} & a_{22} & \cdots & a_{2j} & \cdots & a_{2n} \\ \vdots & \vdots & \cdots & \vdots & \cdots & \vdots \\ a_{i1} & a_{i2} & \cdots & a_{ij} & \cdots & a_{in} \\ \vdots & \vdots & \cdots & \vdots & \cdots & \vdots \\ a_{m1} & a_{m2} & \cdots & a_{mj} & \cdots & a_{nn} \end{bmatrix}$$

It is noted that range of T i.e. $R(T)$, which is also known as range of A, is spanned by column vectors of A i.e. by Set $\{v_1, v_2, ..., v_n\}$, where

$$v_1 = \begin{bmatrix} a_{11} \\ a_{21} \\ \vdots \\ a_{m1} \end{bmatrix}, v_2 = \begin{bmatrix} a_{12} \\ a_{22} \\ \vdots \\ a_{m2} \end{bmatrix}, ..., v_j = \begin{bmatrix} a_{1j} \\ a_{2j} \\ \vdots \\ a_{mj} \end{bmatrix}, ..., v_n = \begin{bmatrix} a_{1n} \\ a_{2n} \\ \vdots \\ a_{mn} \end{bmatrix}.$$

Therefore rank of T = Rank of A = Number of linearly independent column vectors $\{v_1, v_2, ..., v_n\}$ of A.

Definition 4.6: The rank of a matrix A is the maximum number of linearly independent column vectors of A.

This is also known as column rank of matrix A.

Again, we consider vectors from rows of matrix A

$$u_1 = (a_{11}, a_{12}, ..., a_{1n}), u_2 = (a_{21}, a_{22}, ..., a_{2n}) ...$$
$$u_i = (a_{i1}, a_{i2}, ..., a_{in}), ..., u_m = (a_{m1}, a_{m2}, ..., a_{mn}).$$

These m vectors $\{u_1, u_2, ..., u_m\}$ are known as row vectors of matrix A.

Maximum number of linearly independent row vectors of $\{u_1, u_2, ..., u_m\}$ is known as row-rank of matrix A.

Theorem 3.7: Row rank and column rank of a matrix A are equal and each is equal to the rank of the matrix.

Proof: Let a $m \times n$ matrix be

$$A = \begin{bmatrix} a_{11} & a_{12} & \cdots & a_{1j} & \cdots & a_{1n} \\ a_{21} & a_{22} & \cdots & a_{2j} & \cdots & a_{2n} \\ \vdots & \vdots & \cdots & \vdots & \cdots & \vdots \\ a_{i1} & a_{i2} & \cdots & a_{ij} & \cdots & a_{in} \\ \vdots & \vdots & \cdots & \vdots & \cdots & \vdots \\ a_{m1} & a_{m2} & \cdots & a_{mj} & \cdots & a_{nn} \end{bmatrix}.$$

Now applying elementary row operations, suppose above matrix A is reduced to row-reduced echelon form.

Suppose non-zero rows in row-reduced echelon form are $r \leq m$. These r rows are linearly independent. Therefore row rank of $A = r$, since other rows of A can be obtained as linear combination of rows of row-reduced echelon form by considering, the reverse process of above elementary row operations done on A to reduce it to row-reduced echelon form. For example at some step, if an row is divided by a non-zero number, then reverse elementary row operations is multiplication by a that non-zero number.

Therefore we can get matrix A by reverse elementary row operations (including $(m - r)$ additional rows), since rows of A are linear combination of the rows of row-reduced echelon form.

Therefore we conclude that row-rank of a matrix $A = r \leq m$ is equal to the number of non-zero rows in row reduced echelon form.

Therefore we have proved that row rank of matrix A = rank of A.

Now to prove that column rank of matrix A = row-rank of A. Consider the column vectors of matrix A,

$$\begin{bmatrix} a_{11} \\ a_{21} \\ \vdots \\ a_{m1} \end{bmatrix}, \ldots, \begin{bmatrix} a_{1j} \\ a_{2j} \\ \vdots \\ a_{mj} \end{bmatrix}, \ldots, \begin{bmatrix} a_{1n} \\ a_{2n} \\ \vdots \\ a_{mn} \end{bmatrix}.$$

These are n column vectors each of m components i.e., of V_m a vector space of dimension m. Therefore if $n > m$, then these n vectors are linearly dependent in the row space of matrix A. Hence number of linearly independent vectors $r \leq m \leq n$.

If r = row-rank of A = dimension of the row space of A, then $n > r$ and these n vectors are linearly dependent in r-dimensional vector space. Therefore maximum number of linearly independent column vectors $= r$ = row-rank of A.

Hence proved.

EXERCISE SET 4

1. Let $T : U \to V$ be a map from a vector space U to a vector space V. Prove that T is linear if and only if $T(\alpha u_1 + \beta u_2) = \alpha Tu_1 + \beta Tu_2$.

2. Show that $T : V_2 \to V_3$ defined by $T(x_1, x_2) = (0, x_1, x_2) \ \forall \ x_1, x_2 \in V_2$ is a linear map.

3. Show that $T : V_3 \to V_3$ defined by $T(x_1, x_2, x_3) = (x_3 - x_2, x_1, x_2)$ is a linear map.

4. Show that $T : V_3 \to V_3$ defined by $T(x_1, x_2, x_3) = (x_1, x_2^2, x_3)$ is not a linear map.

5. If $T : P_2 \to V_3$, defined by $T(p(x)) = (\alpha, \beta + \gamma, \beta - \gamma)$, where $p(x) = \alpha + \beta x + \gamma x^2$, then show that T is a linear map.

6. Show that a linear transformation from set of real numbers R to R, is a scalar multiple by a fixed real number.

7. Show that transformation $T : V_3 \to P_2$ defined by

    $T(\alpha, \beta, \gamma) = \alpha + \beta + (\beta - \gamma)x + (\alpha + \beta + \gamma)x^2$ is a linear map.

8. Show that $T : V_3 \to P_2$ defined by $T(\alpha, \beta, \gamma) = \alpha + \beta\gamma x + x^2$ is not a linear map.

9. Show that map $T : P_3 \to P_2$ defined by $\dfrac{d}{dx} p(x)$ for all $p(x) \in P_3$ is linear.

10. Find the null space and range space of the linear transformation $T : V_3 \to V_3$, defined by $T(x_1, x_2, x_3) = (x_1 - x_2, x_2 + x_3, x_3 - x_1)$ for $(x_1, x_2, x_3) \in V_3$.

    Also find bases of $N(T)$ and $R(T)$.

11. Let $T : P_3 \to P_2$ be a linear transformation defined by
    $T(a_0 + a_1 x + a_2 x^2 + a_3 x^3) = a_0 + (a_1 + a_2 + a_3)x + (a_1 - a_2 + a_3)x^2$. Find null space and range space of T.

12. Find $n(T)$ and $r(T)$ of the linear map $T : P_3 \to P_2$ defined by

    $T(\alpha_0 + \alpha_1 x + \alpha_2 x^2 + \alpha_3 x^3) = (\alpha_0 - \alpha_1) + (\alpha_1 - \alpha_2)x + \alpha_3 x^2$.

    Is T onto map?

13. Find a basis of $N(T)$ and a basis of $R(T)$ of the linear transformation $T : V_4 \to V_3$ defined by $T(x_1, x_2, x_3, x_n) = (x_1 - x_2, 0, x_3 - x_4)$. Hence find $n(T)$ and $r(T)$.

14. Find bases of $N(T)$ and $R(T)$ of the linear transformation $T : P_2 \to V_3$ defined by

    $T(\alpha + \beta x + \gamma x^2) = (\alpha - 2\beta, \beta - 2\gamma, \alpha - \beta - 2\gamma)$. Hence find $n(T)$ and $r(T)$.

15. Find the bases of $N(T)$ and $R(T)$ of the linear transformation $T : V_3 \to P_2$ defined by $T(x_1, x_2, x_3) = x_1 + (x_2 - x_3)x + (x_1 + x_2 + x_3)x^2$.
    Is T one-one and onto, if so find T^{-1}.

16. Find $N(T)$ and $R(T)$ of the linear transformation $T : P_2 \to V_3$ defined by $T(\alpha + \beta x + \gamma x^2) = (\alpha, \beta - \gamma, \alpha + \gamma)$. Is T one-one, onto, if so find T^{-1}.

17. Check whether linear map $T : P_2 \to P_2$ defined by $T(\alpha + \beta x + \gamma x^2) = (\beta + \gamma) + (\alpha + \gamma)x + (\alpha + \beta)x^2$ is one-one and onto. If so find T^{-1}.

18. Let U be a vector space of dimension n and $T : U \to V$ be a onto linear map. Prove that T is one-one iff dim $V = n$.

19. Find a basis of $N(T)$ and a basis of $R(T)$ of the linear transformation $T : V_3 \to V_3$, defined by $T(x_1, x_2, x_3) = (0, x_1 - 2x_2 - 3x_3, 0)$.
    Hence find $n(T)$ and $r(T)$.

20. Find a non-trivial linear transformation $T : V_3 \to V_3$ such that
    $N(T) = \{(x_1, x_2, x_3) \in V_3 | x_1 = x_2 \text{ and } x_1 + x_2 - x_3 = 0\}$. Find a basis of $N(T)$.

21. Find a non-trivial linear transformation $T : P_2 \to P_2$ such that
    $N(T) = \{p(x) = a + bx + cx^2 \in P_2 | a + 2b + 3c = 0\}$. Find a basis of $N(T)$.

22. Find a non-trivial linear transformation $T : V_3 \to V_3$ such that
    $R(T) = \{(x_1, x_2, x_3) \in V_3 | 2x_1 + x_2 - x_3 = 0\}$. Hence find a basis of $R(T)$.

23. Find a non-trivial linear transformation $T : P_2 \to P_2$ such that
    $R(T) = \{p(x) \in P_2 | P(1) = 0\}$. Find a basis of $R(T)$.

24. Show that the fallowing linear transformation $T : P_2 \to V_3$ defined by
    $T(\alpha_0 + \alpha_1 x + \alpha_2 x^2) = (\alpha_0 + 2\alpha_1 + \alpha_2, -\alpha_0 + \alpha_1, 5\alpha_0 - \alpha_1 + 2\alpha_2)$ is (a) One to one and onto and if so (b) Find T^{-1} inverse of T.

25. Let: $T : P_2 \to P_2$ be a linear transformation defined by
    $T(1) = x - 1, T(x) = x^2 - x, T(x^2) = x^2 + 1$ with standard basis $B = \{1, x, x^2\}$ of P_2, $B = \{1, x, x^2\}$, Find T^{-1}.

26. Let linear maps $T : V_3 \to P_2$ be defined by $T(a, b, c) = c + ax + bx^2$ and $S : V_3 \to P_2$ be defined by $S(a, b, c) = (a - b) + (b - c)x + (c - a)x^2$. Find linear map $(T + S)$.

27. Let $T : V_3 \to P_2$ be defined by $T(a, b, c) = (a - b) + (b - c)x + (c - a)x^2$ and $S : P_2 \to V_3$ be defined by $S(a + bx + cx^2) = (a, b, c)$. Compute composition of linear. Transformations (SoT) and (ToS). Hence check whether $(SoT) \equiv (ToS)$.

28. Let $T : V_4 \to V_4$ be defined by $T(\alpha, \beta, \gamma, \delta) = (0, \alpha, \beta, \gamma)$. Check whether T is nilpotent. If so find the order of nilpotency

29. Write a linear idempotent map T from $P_2 \to P_2$.

30. Let $T : V_3 \to V_3$ be a linear map.

    $B_1 = \{(1, 0, 0), (0, 1, 0), (0, 0, 1)\}$ and

    $B_2 = \{(1, 1, 1), (1, 1, 0), (1, 0, 0)\}$ are two different ordered bases of V_3.

    Matrix associated with linear transformation T is given by

    $$(T : B_1, B_2) = \begin{bmatrix} 1 & 0 & 0 \\ 0 & 1 & 0 \\ 0 & 0 & 1 \end{bmatrix}$$

    Find the expression for $T(x, y, z)$, where (x, y, z) is any element of V_3.

31. Let $T : P_2 \to P_2$ be linear transformation on P_2, space of polynomials of degree ≤ 2, defined by $T(p(x)) = p((x+1))$. Given $B = \left\{1, x, \dfrac{x(x-1)}{2}\right\}$ an order basis for P_2.

    Find the matrix $(T : B, B)$.

32. Let $T : P_2 \to P_2$ be linear transformation on P_2 space of polynomials of degree z_2, defined by $Tf(x) = xf'(x+1)$ Given $B = \{1, x+1, x+x^2\}$ a ordered basis for P_2 find the matrix $(T : B, B)$.

33. For a given matrix $A = \begin{bmatrix} 1 & 1 & 2 & 3 \\ 1 & 0 & 1 & -1 \\ 1 & 2 & 0 & 0 \end{bmatrix}$ and the ordered bases

    $B_1 = \{(1, 1, 1, 2), (1, -1, 0, 0), (0, 0, 1, 1), (0, 1, 0, 0)\}$ of V_4 and

    $B_2 = \{(1, 2, 3), (1, -1, 1), (2, 1, 1)\}$ of V_3.

    Determine a linear transformation $T : V_4 \to V_3$ such that $A = (T : B_1, B_2)$.

34. For the given matrix

    $$A = \begin{bmatrix} 1 & 2 & -2 & 3 & -1 \\ 3 & -1 & 2 & -4 & 5 \\ 2 & 1 & 3 & 2 & 3 \\ -3 & 3 & 4 & 5 & 1 \end{bmatrix}$$

4.36   *Elementary Linear Algebra*

Find the number of linearly independent rows and number of linearly independent columns. Hence verify that both numbers are same.

35. For the given matrix $A = \begin{bmatrix} 2 & 3 & -1 & 4 \\ -1 & 2 & 3 & -2 \\ 3 & 1 & 2 & 4 \\ 1 & -2 & -3 & 1 \\ 4 & 5 & 4 & 3 \end{bmatrix}$

Find the number of linearly independent rows and number of linearly independent columns. Hence verify that both numbers are same.

ANSWERS TO EXERCISE SET – 4

10. $N = \{(0, 0, 0)\}$, $n(T) = V_3$, Basis of $R(T) = \{(1, 0, -1), (-1, 1, 0), (0, 1, 1)\}$, $r(T) = 3$.
11. $N(T) = [(x - x^3)]$, $n(T) = 1$, $R(T) = [\{1, x + x^2, x - x^2\}]$, $r(T) = 3$.
12. $N(T) = [(1 + x + x^3)]$, $n(T) = 1$, $R(T) = [\{1, x, x^2\}]$, $r(T) = 3$.
13. Basis of $N(T) = \{1, 1, 0, 0), (0, 0, 1, 1)\}$ $n(T) = 2$, Basis of $R(T) = \{(1, 0, 0), (0, 01)\}$ $n(T) = 2$.
14. $N(T) = |(4 + 2x + x^3)|$, $n(T) = 1$, $R(T) = \{(2, -1, 1), (0, 1, 1)\}$, $r(T) = 2$.
15. $N(T) = \{0_{V_3}\}$, $R(T) = p_2$, T is one-one onto

    $T^{-1}(\alpha + \beta x + \gamma x^2) = \left(a, \frac{1}{2}(-\alpha+\beta+\gamma), \frac{1}{2}(\gamma-\alpha-\beta)\right)$

16. $N(T) = \{0\}$, $R(T) = V_3$, T is one-one and onto
    $T^{-1}(a, b, c) = (a, (b + c - a), (c - a))$
17. T is one-one onto $T^{-1}(a + bx + cx^2) = \left(\frac{1}{2}(-a+b+c)+\frac{1}{2}(a-b+c)x\frac{1}{2}(a+b-c)x^2\right)$
19. Basis of $N(T) = \{(2, 1, 0), (3, 0, 1)\}$, $n(T) = 2$, Basis of $R(T) = \{(0, 1, 0)\}$, $r(T) = 1$
20. $T(x_1, x_2, x_3) = (x_2 - x_1, x_1 + x_2 - x_3, 0)$, Basis of $N(T) = \{(2, 1, 0), (3, 0, 1)\}$
21. $T(a + bx + cx^2) = a + 2b + 3c + 0x + 0x^2$ Basis of $N(T) = \{x - 2, x^2 - 3\}$
22. $T(\alpha + \beta + \gamma) = (\alpha, \beta, 2\alpha + \beta)$, basis of $R(T) = \{(1, 0, 2), (0, 1, 1)\}$
23. $T(a_0 + a_1x + a_2x_2) = a_0 + a_1x - (a_0 + a_1)x^2$ Basis of $R(T) = \{1 - x^2, x - x^2\}$

24. T is one-one and onto $T^{-1}(a, b, c)$

$= \frac{1}{2}(2a - 5b + c) + \frac{1}{2}(2a - 3b + c)x + \frac{1}{2}(-4a + 11b + 3c)x^2$

25. $T^{-1}(1) = \frac{1}{2}(x^2 - x - 1)$, $T^{-1}(x) = \frac{1}{2}(x^2 - x + 1)$, $T^{-1}(x^2) = \frac{1}{2}(x^2 + x + 1)$

26. $(T + S)(a, b, c) = (a - b + c) + (a + b - c)x + (-a + b + c)x^2$

28. 4

29. $T(a + bx + cx^2) = a + bx$,

30. $T(x, y, z) = (x + y + z, x + y, x)$

31. $\begin{bmatrix} 1 & 0 & -1 \\ 0 & 1 & 1 \\ 0 & 0 & 1 \end{bmatrix}$

32. $\begin{bmatrix} 0 & -1 & 1 \\ 0 & 1 & -1 \\ 0 & 0 & 2 \end{bmatrix}$,

33. $T(x_1, x_2, x_3, x_4) = (-x_2 + x_3 + x_4, \ x_1 + x_2 + x_4, \ x_1 - x_3 + x_4)$

CHAPTER 5

Eigenvalues and Eigenvectors

Various aspects of linear transformations have been discussed in chapter 3. A linear transformation T on the same vector space V i.e. $T: V \to V$ is of more importance, if T maps a vector $x \in V$ to a scalar multiple of x itself i.e. $Tx = \lambda x$ for some suitable values of scalar λ and corresponding vector x. Properties of such scalars and vectors x are discussed in this chapter.

5.1 EIGENVALUES AND EIGENVECTORS

We have studied that a linear transformation on a finite dimensional vector space is represented by a matrix. A linear transformation T from an n-dimensional vector space V to itself would be represented by an $n \times n$ matrix. Therefore, we consider $A_{n \times n}$ square matrix to study the properties of such scalars λ and corresponding vectors x of $Ax = \lambda x$.

Definition 5.1: Given A square matrix of order n, the non-trivial solution vector x of the equation $Ax = \lambda x$ for a suitable scalar value λ is called an eigenvector of matrix A corresponding to λ, which is called an eigenvalue of A.

Set of all eigenvectors corresponding to λ is denoted by $E(\lambda)$, $E(\lambda) \cup \{0\}$ is known as eigenspace of λ.

Note: By definition as above a zero vector is never an eigenvector, but eigenvalue λ may be zero scalar in some cases.

For numerical computation of the eigenvalues and eigenvectors of a given $n \times n$ matrix A, matrix equation $Ax = \lambda x$ can be written as equation $Ax - \lambda Ix = 0$, where I is identity matrix of order n.

The system of linear equation $(A - \lambda I)x = 0$ has non trivial solution if $\det(A - \lambda I) = 0$, i.e. $|A - \lambda I| = 0$, which gives a polynomial equation of order n in λ.

Therefore an $n \times n$ square matrix A has n eigenvalues, which may be all real, some real and some complex or all complex. After computing eigenvalue λ of a given matrix A, $Ax = \lambda x$ can be solved for vector x by the method of system of linear equations.

Following is very well known important result of giving the maximum size of numerical value of the eigenvalues.

Example 1: Find all the eigenvalues and corresponding eigenvectors of the matrix

$$A = \begin{bmatrix} 2 & 2 & 1 \\ -4 & 8 & 1 \\ -1 & -2 & 0 \end{bmatrix}.$$

Let λ be an eigenvalue. Therefore $|A - \lambda I| = 0$

$$\begin{vmatrix} 2-\lambda & 2 & 1 \\ -4 & 8-\lambda & 1 \\ -1 & -2 & -\lambda \end{vmatrix} = 0$$

$$\lambda^3 - 10\lambda^2 + 27\lambda - 18 = 0$$

$$(\lambda - 1)(\lambda^2 - 9\lambda + 18) = 0$$

$$(\lambda - 1)(\lambda - 3)(\lambda - 6) = 0$$

$\Rightarrow \lambda = 1, 3, 6$ all eigenvalues are real and distinct.

Now finding eigenvector, for $\lambda = 1$, $Ax = \lambda x$, gives

$$\begin{bmatrix} 1 & 2 & 1 \\ -4 & 7 & 1 \\ -1 & -2 & -1 \end{bmatrix} \begin{bmatrix} x_1 \\ x_2 \\ x_3 \end{bmatrix} = \begin{bmatrix} 0 \\ 0 \\ 0 \end{bmatrix}$$

$$x_1 + 2x_2 + x_3 = 0$$
$$-4x_1 + 7x_2 + x_3 = 0$$
$$-x_1 - 2x_2 - x_3 = 0$$

Therefore $x_2 = x_1$, $x_3 = -3x_1$.

These three equations reduces to two equations

$$x_1 + 2x_2 + x_3 = 0$$
$$-4x_1 + 7x_2 + x_3 = 0$$

On solving $x_2 = x_1$, $x_3 = -3x_1$.

Therefore $(x_1, x_2, x_3)^T = (x_1, x_1, -3x_1)^T = (x_1, x_1, -3x_1)^T$, $x_1 \neq 0$

$$E(1) = \left[(1, 1, -3)^T\right] - (0, 0, 0).$$

For $\lambda = 3$

$$\begin{bmatrix} -1 & 2 & 1 \\ -4 & 5 & 1 \\ -1 & -2 & -3 \end{bmatrix} \begin{bmatrix} x_1 \\ x_2 \\ x_3 \end{bmatrix} = \begin{bmatrix} 0 \\ 0 \\ 0 \end{bmatrix}$$

$$-x_1 + 2x_2 + x_3 = 0$$
$$-4x_1 + 5x_2 + x_3 = 0$$
$$-x_1 - 2x_2 - 3x_3 = 0.$$

On solving these equations,

we get
$$x_2 = -x_3, \ x_1 = -x_3.$$

Therefore eigenvector $(x_1, x_2, x_3)^T = (-x_3, -x_3, x_3)^T = -x_3(1, 1, -1)^T$, $x_3 \neq 0$

$$E(3) = \left[(1, 1, -1)^T\right] - (0, 0, 0).$$

For $\lambda = 6$, system of equations is

$$\begin{bmatrix} -4 & 2 & 1 \\ -4 & 2 & 1 \\ -1 & -2 & -6 \end{bmatrix} \begin{bmatrix} x_1 \\ x_2 \\ x_3 \end{bmatrix} = \begin{bmatrix} 0 \\ 0 \\ 0 \end{bmatrix}.$$

These equations are equivalent to

$$-4x_1 + 2x_2 + x_3 = 0$$
$$-x_1 - 2x_2 - 6x_3 = 0.$$

On adding $-5x_1 - 5x_3 = 0$, $-5x_1 - 5x_3 = 0$.

On substituting in the above equation

$$-x_1 - 2x_2 - 6x_1 = 0$$

$$-2x_2 + 5x_1 = 0 \ \ x_2 = \frac{5}{2} x_1$$

Therefore eigenvector

$$(x_1, x_2, x_3)^T = \left(x_1, \frac{5}{2} x_1, -x_1\right)^T = \frac{1}{2} x_1 (2, 5, -2)^T, \ x_1 \neq 0$$

$E(6) = \left[(2, 5, -2)^T\right] - (0, 0, 0)$. Eigenspace of an eigenvalue 6 is $\left[(2, 5, -2)^T\right]$.

Example 2: Find all the eigenvalues and corresponding eigenvectors of the matrix

$$A = \begin{bmatrix} 1 & 2 & 1 \\ 6 & -1 & 0 \\ -1 & -2 & -1 \end{bmatrix}.$$

Characteristic equation is $\begin{vmatrix} 1-\lambda & 2 & 1 \\ 6 & -1-\lambda & 0 \\ -1 & -2 & -1-\lambda \end{vmatrix} = 0.$

On solving $\lambda^3 + \lambda^2 - 12\lambda = 0$, eigenvalues are $\lambda = -4, 0, 3$.

For eigenvector corresponding to, $\lambda = -4$, equations are

$$5x_1 + 2x_2 + x_3 = 0$$
$$6x_1 + 3x_2 = 0$$
$$-x_1 - 2x_2 + 3x_3 = 0.$$

On solving the above equations, we get $x_2 = -2x_1$, $x_3 = -x_1$, therefore eigenvector is $(x_1, -2x_1, -x_1)^T$, $x_1 \neq 0$.

Therefore eigenvectors are $[(1, -2, -1)]^T - (0, 0, 0)$.

For eigenvector corresponding to, $\lambda = 0$, equations are

$$x_1 + 2x_2 + x_3 = 0$$
$$6x_1 - x_2 = 0$$
$$-x_1 - 2x_2 - x_3 = 0.$$

On solving the above equations, we get $x_2 = 6x_1$, $x_3 = -13x_1$, therefore eigenvector is $(x_1, 6x_1, -13x_1)^T$, $x_1 \neq 0$.

Eigenvectors are $[(1, 6, -13)]^T - (0, 0, 0)$.

For eigenvector corresponding to, $\lambda = 3$, equations are

$$-2x_1 + 2x_2 + x_3 = 0$$
$$6x_1 - 4x_2 = 0$$
$$-x_1 - 2x_2 - 4x_3 = 0.$$

On solving the above equations, we get $2x_2 = 3x_1$, $x_3 = -x_1$, therefore eigenvector is $(2x_1, 3x_1, -2x_1)^T$, $x_1 \neq 0$.

Eigenvectors are $[(2, 3, -2)]^T - (0, 0, 0)$.

Example 3: Find all the eigenvalues and corresponding eigenvectors of the matrix

$$A = \begin{bmatrix} 3 & 2 & 4 \\ 2 & 0 & 2 \\ 4 & 2 & 3 \end{bmatrix}.$$

Characteristic equation is det $(A - \lambda I) = 0$. Det $\begin{bmatrix} 3-\lambda & 2 & 4 \\ 2 & -\lambda & 2 \\ 4 & 2 & 3-\lambda \end{bmatrix} = 0$

i.e., $\lambda^3 - 6\lambda^2 - 15\lambda - 8 = 0$, $(\lambda + 1)(\lambda^2 - 7\lambda - 8) = 0$

$\lambda = 8, -1, -1.$

Eigenvector corresponding to eigenvalue $\lambda = 8$, $(A - 8I)x = 0$, where $x = [x_1, x_2, x_3]^T$.

$$\begin{bmatrix} -5 & 2 & 4 \\ 2 & -8 & 2 \\ 4 & 2 & -5 \end{bmatrix} \begin{bmatrix} x_1 \\ x_2 \\ x_3 \end{bmatrix} = \begin{bmatrix} 0 \\ 0 \\ 0 \end{bmatrix}$$

$$-5x_1 + 2x_2 + 4x_3 = 0$$
$$2x_1 - 8x_2 + 2x_3 = 0$$
$$-4x_1 + 2x_2 - 5x_3 = 0$$

On solving these equations, we get

$$x_1 = x_3, \, x_1 = 2x_2.$$

Therefore $x = (x_1, x_2, x_3)^T = \left(x_1, \dfrac{x_1}{2}, x_1\right)^T = \dfrac{x_1}{2}(2, 1, 2), \, x_1 \neq 0$.

Eigenvectors are $[(2, 1, 2)] - (0, 0, 0)$ corresponding to eigenvalues $\lambda = 8$.
For $\lambda = -1$, equations become

$$4x_1 + 2x_2 + 4x_3 = 0$$
$$2x_1 + x_2 + 2x_3 = 0$$
$$4x_1 + 2x_2 + 4x_3 = 0.$$

All these equations are equivalent to a single equation

$$2x_1 + x_2 + 2x_3 = 0.$$

Therefore eigenvector$(x_1, x_2, x_3)^T = (x_1, -2x_1 - 2x_3, x_3)$, on replacing x_2 by $-2x_1 - 2x_3$,

Is $\quad (x_1, -2x_1, 0) + (0, -2x_3, x_3) = x_1(1, -2, 0) + x_3(0, -2, 1)$

$\hspace{5cm} = x_1(1, -2, 0) + x_3(0, -2, 1), \ |x_1| + |x_3| \neq 0$.

Hence set of eigenvectors corresponding to $\lambda = -1$

$\hspace{5cm} = [(1, -2, 0), (0, -2, 1)] - (0, 0, 0)$.

Space of eigenvectors corresponding to $\lambda = -1$ is $[(1, -2, 0), (0, -2, 1)]$.

Note: This is two dimensional vector space.

Example 4: Find all the eigenvalues and corresponding eigenvectors of the matrix

$$A = \begin{bmatrix} 1 & 2 & -2 \\ 2 & 1 & 2 \\ -2 & 2 & 1 \end{bmatrix}.$$

Characteristic equation $\det(A - \lambda I) = 0$, $\det \begin{bmatrix} 1-\lambda & 2 & -2 \\ 2 & 1-\lambda & 2 \\ -2 & 2 & 1-\lambda \end{bmatrix} = 0$

Is $\quad \lambda^3 - 3\lambda^2 - 9\lambda + 27 = 0$, $(\lambda - 3)(\lambda^2 - 9) = 0$.

Eigenvalues are $\lambda = 3, 3, -3$.

For $\lambda = -3$, equations for eigenvectors are

$$\begin{bmatrix} 4 & 2 & -2 \\ 2 & 4 & 2 \\ -2 & 2 & 4 \end{bmatrix} \begin{bmatrix} x_1 \\ x_2 \\ x_3 \end{bmatrix} = \begin{bmatrix} 0 \\ 0 \\ 0 \end{bmatrix} \Rightarrow$$

$$4x_1 + 2x_2 - 2x_3 = 0$$
$$2x_1 + 4x_2 + 2x_3 = 0$$
$$-2x_1 + 2x_2 + 4x_3 = 0.$$

On solving these equations, we get $x_2 = -x_1$, $x_3 = x_1$.

Therefore eigenvectors are $(x_1, -x_1, x_1)^T$, $x_1 \neq 0$, i.e., $[(1, -1, 1)]^T - (0, 0, 0)$.

For $\lambda = 3$, equations for eigenvectors are

Eigenvalues and Eigenvectors

$$\begin{bmatrix} -2 & 2 & -2 \\ 2 & -2 & 2 \\ -2 & 2 & -2 \end{bmatrix} \begin{bmatrix} x_1 \\ x_2 \\ x_3 \end{bmatrix} = \begin{bmatrix} 0 \\ 0 \\ 0 \end{bmatrix} \Rightarrow$$

$$-2x_1 + 2x_2 - 2x_3 = 0$$
$$2x_1 - 2x_2 + 2x_3 = 0$$
$$-2x_1 + 2x_2 + -2x_3 = 0.$$

On solving these equations, we get $x_2 = x_3 + x_1$.

Therefore eigenvectors are $(x_1, x_3 + x_1, x_3)^T = ((x_1, x_1, 0)^T + (0, x_3, x_3))^T$.

Eigenvectors are $[\{(1, 1, 0), (0, 1, 1)\}]^T - (0, 0, 0)$.

In the above two examples, we have computed eigenvalues and corresponding eigenvectors. We would like to know the dependence or independence of the eigenvectors. For this, we have the following theorem.

Theorem 5.1: Let $\{v_1, v_2, ..., v_k\}$ be the set of eigenvectors corresponding to distinct eigenvalues $\lambda_1, \lambda_2, ..., \lambda_k$ of an $n \times n$ square matrix A with $k \leq n$. Then the set $\{v_1, ..., v_k\}$ is linearly independent.

Proof: We shall prove this theorem by induction $v_1 \neq 0$ by definition of an eigenvector, it is a non-zero vector, therefore $\{v_1\}$ is LI.

The theorem is true for one vector v_1,

Now suppose $\{v_1, v_2, ..., v_r\}$ be linearly independent for $r(r < k)$ vectors.

Now, we shall prove that $(r + 1)$ vectors $\{v_1, v_2, ..., v_r, v_{r+1}\}$ are linearly independent.

Consider $\quad c_1 v_1 + c_2 v_2 + , ..., c_r v_r + c_{r+1} v_{r+1} = \mathbf{0}_V$ \hfill (1)

$\mathbf{0}_V$ is used for zero vector.

Applying matrix A on both sides, we have

$$A\left(c_1 v_1 + c_2 v_2 + , ..., + c_r v_r + c_{r+1} v_{r+1}\right) = A\mathbf{0}_V$$

$c_1 A v_1 + c_2 A v_2 + , ..., + c_r A v_r + c_{r+1} A v_{r+1} = \mathbf{0}_V$, since A is linear.

$c_1 \lambda_1 v_1 + c_2 \lambda_2 v_2 + , ..., + c_r \lambda_r v_r + , ..., + c_{r+1} \lambda_{r+1} v_{r+1} = \mathbf{0}_V,$

since $A v_i = \lambda_i v_i$, for $i = 1, 2, ..., r, r + 1$ \hfill (2)

\hfill (3)

Now multiplying (1) by λ_{r+1}, we get

$$c_1 \lambda_{r+1} v_1 + c_2 \lambda_{r+1} v_2 + , ..., + c_r \lambda_{r+1} v_r + c_{r+1} \lambda_{r+1} v_{r+1} = \mathbf{0}_V.$$

On subtracting (3) from (2), we get

$$c_1(\lambda_1 - \lambda_{r+1}) v_1 + c_2(\lambda_2 - \lambda_{r+1}) v_2 + c_3(\lambda_3 - \lambda_{r+1}) v_3, ..., + c_r(\lambda_r - \lambda_{r+1}) v_r = \mathbf{0}_V.$$

Now set $\{v_1, v_2, ..., v_r\}$ is LI,
Therefore
$$c_1(\lambda_1 - \lambda_{r+1}) = 0, c_2(\lambda_2 - \lambda_{r+1}) = 0, c_r(\lambda_r - \lambda_{r+1}) = 0$$

Since $\lambda_1, \lambda_2, ..., \lambda_r, \lambda_{r+1}$ are distinct, therefore $\lambda_i - \lambda_{r+1} \neq 0$, for $i = 1, 2, ..., r$. hence
$$c_1 = c_2 = ... = c_r = 0.$$

Now (1) reduces to $0v_1 + 0v_2 + ... + 0v_r + c_{r+1}v_{r+1} = 0_V$,
$c_{r+1}v_{r+1} = 0 \Rightarrow c_{r+1} = 0$ since $v_{r+1} \neq 0$ being an eigenvector.
Now in (1) all scalars $c_1 = c_2 = c_3 = , ..., = c_r = c_{r+1} = 0$.
Therefore the set $\{v_1, v_2, ..., v_r, v_{r+1}\}$ is linearly independent.

By induction, if true for $r = 1$, since $\{v_1\}$ is linearly independent, true for $r = 2$, $\{v_1, v_2\}$ is linearly independent.

If true for $r = 2$, since $\{v_1, v_2\}$ is linearly independent, then true for $r = 3$, $\{v_1, v_2, v_3\}$ is linearly independent. Continuing by induction, we get $\{v_1, v_2, ..., v_k\}$ is linearly independent.

Certain properties of eigenvalues and eigenvectors are given below:

Consider an $n \times n$ matrix A, and $Ax = \lambda x$, where x is eigenvector corresponding to eigenvalue λ.

Applying matrix A on both sides, $A(Ax) = A(\lambda x) = \lambda Ax \Rightarrow A^2 x = \lambda^2 x$.
Continuing similarly in general, we get $A^k x = \lambda^k x$.

i.e. if λ is an eigenvalue of A, then λ^k is eigenvaolue of A^k and eigenvector x remains the same.

By using property of the eigenvalues, we also get
$$(a_0 A^n + a_1 A^{n-1} + ,..., + a_{n-1} A + a_n I)x = (a_0 \lambda^n + a_1 \lambda^{n-1} + ,..., + a_n)x.$$

Therefore, $(a_0 \lambda^n + a_1 \lambda^{n-1} + ,..., + a_n)$ is eigenvalue of the matrix $(a_0 A^n + ,..., + a_n I)$, with same eigenvector x.

If A is non-singular, then no eigenvalue is zero, In such a case A^{-1} exists and
$$A^{-1} Ax = A^{-1} \lambda x \text{ i.e. } Ix = A^{-1} \lambda x \Rightarrow \frac{1}{\lambda} x = A^{-1} x.$$

This shows that i.e. if λ, x are eigenvalue and corresponding eigenvector of A i.e., $Ax = \lambda x$, then $\frac{1}{\lambda}$ is an eigenvalue of A^{-1} and eigenvector x is unchanged.

Let $Ax = \lambda x$. If eigenvalues and corresponding eigenvectors of transpose of A are to be considered, then it is known that $\det(A - \lambda I) = \det(A - \lambda I)^T = \det(A^T - \lambda I) = 0$.

Therefore eigenvalues of matrix A and eigenvalues of it's transpose matrix A^T are same $(Ax)^T = \lambda(x)^T$ does not give any relation for eigenvectors. Hence eigenvectors of matrix A and it's transpose A^T may not be same in general.

5.2 GERSHGORIN CIRCLE THEOREM

The Gershgorin circle theorem identifies a region in the complex plane that contains all the eigenvalues of an $n \times n$ square matrix $A = (a_{ij})$ (may be matrix of complex numbers also). Then each eigenvalue of A is in at least one of the disks

$$\{z - a_{ii}| < R_I\}, \ i = 1, 2, \ldots, n, \text{ where } R_i = \sum_{\substack{j=1 \\ j \neq 1}}^{n} |a_{ij}|$$

Theorem 5.2: Kayley-Hamilton Theorem. Every square matrix satisfies its own characteristic equation.

Let square matrix be $A = \begin{bmatrix} a_{11} & a_{12} & \cdots & a_{1,n-1} & a_{1,n} \\ a_{21} & a_{22} & \cdots & a_{2,n-1} & a_{2,n} \\ \vdots & \vdots & \cdots & \vdots & \vdots \\ a_{n-1,1} & a_{n-1,1} & \cdots & a_{n-1,n} & a_{n-1n} \\ a_{n1} & a_{n2} & \cdots & a_{n,n-1} & a_{nn} \end{bmatrix}$

$[A - \lambda I] = \begin{bmatrix} a_{11} - \lambda & a_{12} & \cdots & a_{1,n-1} & a_{1,n} \\ a_{21} & a_{22} - \lambda & \cdots & a_{2,n-1} & a_{2,n} \\ \vdots & \vdots & \cdots & \vdots & \vdots \\ a_{n-1,1} & a_{n-1,1} & \cdots & a_{n-1,n1} - \lambda & a_{n-1,n} \\ a_{n1} & a_{n2} & \cdots & a_{n,n-1} & a_{nn} - \lambda \end{bmatrix}$

$|A - \lambda I| = (-1)^n (\lambda^n + a_1 \lambda^{n-1} + a_2 \lambda^{n-2} + \ldots + a_{n-1} \lambda + a_n) = 0$ \hfill (a)

The elements of matrix $A - \lambda I$ are of fist degree in λ so the elements of Adj $(A-\lambda I)$ are polynomial in λ of degree at most $(n-1)$, because some terms may cancelled. Therefore we can write Adj $(A - \lambda I) = B_0 \lambda^{n-1} + B_1 \lambda^{n-2} + \ldots + B_{n-2} \lambda + B_{n-1}$ \hfill (b)

where $B_0, B_1, \ldots B_{n-2}, B_{n-1}$ are matrices of order $n \times n$.

Adj A = Adjoint matrix of $A = (A_{ij})'_{n \times n}$ where, A_{ij} is value of the determinant obtained by deleting from i^{th} row and J^{th} column of matrix A.

Also $A(AdjA) + |A|I = (AdjA)A. = |A| \ I$, where I is identity matrix. \hfill (c)

On substituting from (a) and (b) in (c), we get

$(A-\lambda I)(B_0\lambda^{n-1} + B_1\lambda^{n-2} +...+ B_{n-2}\lambda + B_{n-1})$

$=(-1)^n(\lambda^n + a_1\lambda^{n-1} + a_2\lambda^{n-2} +...+ a_{n-1}\lambda + an)I$. Now equating the coefficient of like powers of λ on both sides, we get

$$-IB_0 = (-1)^n I$$
$$AB_0 - IB_1 = (-1)^n a_1 I$$
$$AB_1 - IB_2 = (1-)^n a_2 I$$
$$\cdots\cdots$$
$$AB_{n-1} = (-1)^n a_n I$$

Premultiplying the above relations by $A^n, A^{n-1}, ..., I$ respectively and adding we get $0 = (-1)^n(A^n + a_1 A^{n-1} + a_2 A^{n-2} + ... a_n I)$.

Hence $A^n + a_1 A^{n-1} + a_2 A^{n-2} +...+ a_n I = 0$.

Inverse of the matrix can be found using **Kayley-Hamilton Theorem as**

Let $|A-\lambda I| = (a_n\lambda^n + a_1\lambda^{n-1} + a_2\lambda^{n-2} +...+ a_{n-1}\lambda + a_n) = 0$ be the characteristic equation of the matrix A.

So we have $a_0 A^n + a_1 A^{n-1} +...+ a_n I = 0$.

Multiplying by A^{-1}, we have $a_0 A^{n-1} + a_1 A^{n-2} + a_2 A^{n-3} +...+ a_n A^{-1} = 0$.

Therefore $A^{-1} = -\dfrac{1}{a_n}(a_0 A^{n-1} + a_1 A^{n-2} + a_2 A^{n-3} +...+ a_{n-1} I)$

Example 5. Find the inverse of the matrix by Kayley-Hamilton Theorem

$$A = \begin{bmatrix} 3 & 2 & 4 \\ 2 & 0 & 2 \\ 4 & 2 & 3 \end{bmatrix}.$$

Solution. Characteristic equation is $\det(A-\lambda I) = \det\begin{bmatrix} 3-\lambda & 2 & 4 \\ 2 & -\lambda & 2 \\ 4 & 2 & 3-\lambda \end{bmatrix} = 0$

i.e., $\lambda^3 - 6\lambda^2 - 15\lambda - 8 = 0$,

Using Kayley-Hamilton Theorem, $A^3 - 6A^2 - 15A - 8I = 0$

$A^{-1} = \dfrac{1}{8}(A^2 - 6A - 15I)$ (Substituting for A) and $A^2 = \begin{bmatrix} 29 & 14 & 28 \\ 14 & 8 & 14 \\ 28 & 14 & 29 \end{bmatrix}$.

$$A^{-1} = \frac{1}{8} \left(\begin{bmatrix} 29 & 14 & 28 \\ 14 & 8 & 14 \\ 28 & 14 & 29 \end{bmatrix} - \begin{bmatrix} 18 & 12 & 24 \\ 12 & 0 & 12 \\ 24 & 12 & 18 \end{bmatrix} - \begin{bmatrix} 15 & 0 & 0 \\ 0 & 15 & 0 \\ 0 & 0 & 15 \end{bmatrix} \right)$$

$$A^{-1} = \begin{bmatrix} -\frac{4}{8} & \frac{2}{8} & \frac{4}{8} \\ \frac{2}{8} & -\frac{7}{8} & \frac{2}{8} \\ \frac{4}{8} & \frac{2}{8} & -\frac{4}{8} \end{bmatrix}$$

Example 6. Find the characteristic equation of the matrix $A = \begin{bmatrix} 2 & 1 & 1 \\ 0 & 1 & 0 \\ 1 & 1 & 2 \end{bmatrix}$, and hence compute A^{-1}.

Solution. Characteristic equation is $\begin{vmatrix} 2-\lambda & 1 & 1 \\ 0 & 1-\lambda & 0 \\ 1 & 1 & 2-\lambda \end{vmatrix} = 0$

$\lambda^3 - 5\lambda^2 + 7\lambda - 3 = 0$.

By Kayley Hamilton theorem

$A^3 - 5A^2 + 7A - 3I = 0$, $A^2 - 5A + 7I - 3A^{-1} = 0$

$$A^{-1} = \frac{1}{3}(A^2 - 5A + 7I) = \frac{1}{3} \left\{ \begin{bmatrix} 5 & 4 & 4 \\ 0 & 1 & 0 \\ 1 & 1 & 2 \end{bmatrix} - 5\begin{bmatrix} 2 & 1 & 1 \\ 0 & 1 & 0 \\ 1 & 1 & 2 \end{bmatrix} + 7\begin{bmatrix} 1 & 0 & 0 \\ 0 & 1 & 0 \\ 0 & 0 & 1 \end{bmatrix} \right\}$$

$$A^{-1} = \frac{1}{3} \left\{ \begin{bmatrix} 5 & 4 & 4 \\ 0 & 1 & 0 \\ 4 & 4 & 5 \end{bmatrix} - \begin{bmatrix} 10 & 5 & 5 \\ 0 & 5 & 0 \\ 5 & 5 & 10 \end{bmatrix} + \begin{bmatrix} 7 & 0 & 0 \\ 0 & 7 & 0 \\ 0 & 0 & 7 \end{bmatrix} \right\}$$

$$A^{-1} = \frac{1}{3} \begin{bmatrix} 2 & -1 & -1 \\ 0 & 3 & 0 \\ -1 & -1 & 2 \end{bmatrix}.$$

Example 7. Find the characteristic equation of the matrix $A = \begin{bmatrix} 2 & -1 & 1 \\ -1 & 2 & -1 \\ 1 & -2 & 2 \end{bmatrix}$,

and hence find the inverse A^{-1}.

Solution. Characteristic equation $\begin{vmatrix} 2-\lambda & -1 & 1 \\ -1 & 2-\lambda & -1 \\ 1 & -2 & 2-\lambda \end{vmatrix} = 0$

$(2-\lambda)\begin{vmatrix} 2-\lambda & -1 \\ -2 & 2-\lambda \end{vmatrix} + \begin{vmatrix} -1 & -1 \\ 1 & 2-\lambda \end{vmatrix} + \begin{vmatrix} -1 & 2-\lambda \\ 1 & -2 \end{vmatrix} = 0$

$(2-\lambda)\{(2-\lambda)^2 - 2\} - (1-\lambda) + \lambda = 0$.

characteristic equation $\lambda^3 - 6\lambda^2 + 8\lambda - 3 = 0$

Using Kayley Hamilton Theorem

$A^{-1} = \dfrac{1}{3}(A^2 - 6A + 8I)$

$A^{-1} = \dfrac{1}{3}\left(\begin{bmatrix} 6 & -6 & 5 \\ -5 & 7 & -5 \\ 6 & -9 & 7 \end{bmatrix} - 6\begin{bmatrix} 2 & -1 & 1 \\ -1 & 2 & -1 \\ 1 & -2 & 2 \end{bmatrix} + 8\begin{bmatrix} 1 & 0 & 0 \\ 0 & 1 & 0 \\ 0 & 0 & 1 \end{bmatrix}\right)$

$A^{-1} = \dfrac{1}{3}\left(\begin{bmatrix} 6 & -6 & 5 \\ -5 & 7 & -5 \\ 6 & -9 & 7 \end{bmatrix} + \begin{bmatrix} -12 & 6 & -6 \\ 6 & -12 & 6 \\ -6 & 12 & -12 \end{bmatrix} + \begin{bmatrix} 8 & 0 & 0 \\ 0 & 8 & 0 \\ 0 & 0 & 8 \end{bmatrix}\right)$

Therefore inverse of the matrix $A^{-1} = \dfrac{1}{3}\begin{bmatrix} 2 & 0 & -1 \\ 1 & 3 & 1 \\ 0 & 3 & 3 \end{bmatrix}$.

Example 8. Find inverse of the matrix A using Kayley–Hamilton theorem, given the matrix

$A = \begin{bmatrix} 2 & -1 & 1 \\ -1 & 2 & -1 \\ 1 & -1 & 2 \end{bmatrix}$.

Eigenvalues and Eigenvectors

5.13

Solution. Characteristic equation $\begin{vmatrix} 2-\lambda & -1 & 1 \\ -1 & 2-\lambda & -1 \\ 1 & -1 & 2-\lambda \end{vmatrix} = 0.$

$(2-\lambda)\begin{vmatrix} 2-\lambda & -1 \\ -1 & 2-\lambda \end{vmatrix} + \begin{vmatrix} -1 & -1 \\ 1 & 2-\lambda \end{vmatrix} + \begin{vmatrix} -1 & 2-\lambda \\ 1 & -1 \end{vmatrix} = 0$

$(2-\lambda)(\lambda^2 - 4\lambda + 3) + (\lambda - 1) + (\lambda - 1) = 0$

$\lambda^3 - 6\lambda^2 + 9\lambda - 4 = 0.$

Using Kayley-Hamilton Theorem $A^{-1} = \dfrac{1}{4}(A^2 - 6A + 9I)$

$A^{-1} = \dfrac{1}{4}\left(\begin{bmatrix} 6 & -5 & 5 \\ -5 & 6 & -5 \\ 5 & -5 & 6 \end{bmatrix} - 6\begin{bmatrix} 2 & -1 & 1 \\ -1 & 2 & -1 \\ 1 & -1 & 2 \end{bmatrix} + 9\begin{bmatrix} 1 & 0 & 0 \\ 0 & 1 & 0 \\ 0 & 0 & 1 \end{bmatrix} \right)$

$A^{-1} = \dfrac{1}{4}\left(\begin{bmatrix} 6 & -5 & 5 \\ -5 & -6 & -5 \\ 5 & -5 & 6 \end{bmatrix} + \begin{bmatrix} -12 & 6 & -6 \\ 6 & -12 & 6 \\ -6 & 6 & -12 \end{bmatrix} + \begin{bmatrix} 9 & 0 & 0 \\ 0 & 9 & 0 \\ 0 & 0 & 9 \end{bmatrix} \right)$

$A^{-1} = \dfrac{1}{4}\begin{bmatrix} 3 & 1 & -1 \\ 1 & 3 & 1 \\ -1 & 1 & 3 \end{bmatrix}$

Example 9. Find the inverse of the matrix by Kayley-Hamilton Theorem

$A = \begin{bmatrix} 1 & 2 & -2 \\ 2 & 1 & 2 \\ -2 & 2 & 1 \end{bmatrix}.$

Solution. Characteristic equation $\det(A - \lambda I) = 0$, $\det \begin{bmatrix} 1-\lambda & 2 & -2 \\ 2 & 1-\lambda & 2 \\ -2 & 2 & 1-\lambda \end{bmatrix} = 0$

$\lambda^3 - 3\lambda^2 - 9\lambda + 27 = 0,$

Using Kayley-Hamilton Theorem, $A^3 - 3A^2 - 9A + 27I = 0,$

$$A^{-1} = \frac{1}{27}(-A^2 + 3A + 9I)$$

$$A^{-1} = \frac{1}{27}\left(-\begin{bmatrix} 9 & 0 & 0 \\ 0 & 9 & 0 \\ 0 & 0 & 9 \end{bmatrix} + 3\begin{bmatrix} 1 & 2 & -2 \\ 2 & 1 & 2 \\ -2 & 2 & 1 \end{bmatrix} + 9\begin{bmatrix} 1 & 0 & 0 \\ 0 & 1 & 0 \\ 0 & 0 & 1 \end{bmatrix}\right)$$

$$A^{-1} = \begin{bmatrix} \frac{1}{9} & \frac{2}{9} & -\frac{2}{9} \\ \frac{2}{9} & \frac{1}{9} & \frac{2}{9} \\ -\frac{2}{9} & \frac{2}{9} & \frac{1}{9} \end{bmatrix}$$

5.3 DIAGONALIZATION OF A MATRIX

Definition 5.2: If A and B are two $n \times n$ matrices, B and A are called similar to each other, if there exists an $n \times n$ invertible matrix P such that

$$PAP^{-1} = B, \text{ which is also same as } A = P^{-1}BP,$$

because P and P^{-1} are inverse of each other, therefore any one of these two can be named P and other P^{-1}.

If in the above relation, B is a diagonal matrix then matrix P is called similarity transformation. Finding such a matrix B is called diagonalization of A.

Further similarity transformation retains same eigenvalues but changed eigenvectors i.e. matrices A and B will have same eigenvalues but eigenvectors may not remain the same.

Theorem 5.3: An $n \times n$ matrix A is diagonalizable if and only if A has n linearly independent eigenvectors, which form the basis for P i.e. similarity transformation.

Proof: We can construct a matrix P, whose columns are linearly independent vectors, and $P^{-1}AP = D$, where D is a diagonal matrix.

Consider matrix $P = [v_1, v_2, ..., v_n]$, where $\{v_1, v_2, ..., v_n\}$ are n linearly independent eigenvectors of A. v_i is written as column vector.

Now $\quad AP = A[v_1, v_2, ..., v_n] = [Av_1, Av_2, ..., Av_n]$

and $\quad PD = [v_1, v_2, ..., v_n]\begin{bmatrix} \lambda_1 & 0 & & 0 \\ 0 & \lambda_2 & & 0 \\ 0 & 0 & & 0 \\ 0 & 0 & 0 & \lambda_n \end{bmatrix}$

Eigenvalues and Eigenvectors

$$= [\lambda_1 v_1, \lambda_2 v_2, ..., \lambda_n v_n]$$

Now, $Av_i = \lambda_i v_i$ for $i = 1, 2, ..., n$ since $\{v_1, v_2, ..., v_n\}$ are eigenvectors of A.

From above two relations we get $AP = PD$, therefore $P^{-1}AP = D$, A is diagonalizable.

And if A is diagonalizable i.e. $P^{-1}AP = D$, then

$$AP = PD \text{ i.e. } [Av_1, Av_2, ..., Av_n] = [\lambda_1 v_1, ..., \lambda_n v_n].$$

On equating the components, we ge $Av_i = \lambda_i v_i$, $i = 1, 2, ..., n$.

Example 10: Consider the matrix $A = \begin{bmatrix} 0 & 2 & 1 \\ -4 & 6 & 1 \\ -1 & -2 & -2 \end{bmatrix}$ which has eigenvalues $\lambda = 4, -1,$

1 with corresponding eigenvectors $(2, 5, -2)^T$, $(1, 1, -3)^T$ and $(1, 1, -1)^T$. Eigenvalues are distinct therefore eigenvectors are linearly independent.

After diagonalization, we must get

$$P^{-1}AP = \begin{bmatrix} 4 & 0 & 0 \\ 0 & -1 & 0 \\ 0 & 0 & 1 \end{bmatrix},$$ diagonal entries are eigenvalues in that order.

Now to check this $P = \begin{bmatrix} 2 & 1 & 1 \\ 5 & 1 & 1 \\ -2 & -3 & -1 \end{bmatrix}$, and $P^{-1} = \begin{bmatrix} -\frac{1}{3} & \frac{1}{3} & 0 \\ -\frac{1}{2} & 0 & -\frac{1}{2} \\ \frac{13}{6} & -\frac{2}{3} & \frac{1}{2} \end{bmatrix}$

Now $\quad AP = \begin{bmatrix} 0 & 2 & 1 \\ -4 & 6 & 1 \\ -1 & -2 & -2 \end{bmatrix} \begin{bmatrix} 2 & 1 & 1 \\ 5 & 1 & 1 \\ -2 & -3 & -1 \end{bmatrix} = \begin{bmatrix} 8 & -1 & 1 \\ 20 & -1 & 1 \\ -8 & 3 & -1 \end{bmatrix}$

and $\quad P^{-1}AP = \begin{bmatrix} -\frac{1}{3} & \frac{1}{3} & 0 \\ -\frac{1}{2} & 0 & -\frac{1}{2} \\ \frac{13}{6} & -\frac{2}{3} & \frac{1}{2} \end{bmatrix} \begin{bmatrix} 8 & -1 & 1 \\ 20 & -1 & 1 \\ -8 & 3 & -1 \end{bmatrix}.$

$$= \begin{bmatrix} 4 & 0 & 0 \\ 0 & -1 & 0 \\ 0 & 0 & 1 \end{bmatrix}$$

The result is verified.

Example 11: Diagonalize the given matrix A,

$$A = \begin{bmatrix} 5 & -6 & -6 \\ -1 & 4 & 2 \\ 3 & -6 & -4 \end{bmatrix}$$

with eigenvalues 1, 2, 2, and with corresponding eigenvectors $(3, -1, 3)$, $(2, 1, 0)$, and $(2, 0, 1)$.

Above three vectors $\{(3, -1, 3), (2, 1, 0), (2, 0, 1)\}$ are linearly independent. Matrix P of eigenvector is

$$P = \begin{bmatrix} 3 & 2 & 2 \\ -1 & 1 & 0 \\ 3 & 0 & 1 \end{bmatrix} \text{ and it's inverse. } P^{-1} = \begin{bmatrix} -1 & 2 & 2 \\ -1 & 3 & 2 \\ 3 & -6 & -5 \end{bmatrix}$$

Now $$AP = \begin{bmatrix} 5 & -6 & -6 \\ -1 & 4 & 2 \\ 3 & -6 & -4 \end{bmatrix} \begin{bmatrix} 3 & 2 & 2 \\ -1 & 1 & 0 \\ 3 & 0 & 1 \end{bmatrix} = \begin{bmatrix} 3 & 4 & 4 \\ -1 & 2 & 0 \\ 3 & 0 & 2 \end{bmatrix}.$$

and $$P^{-1}AP = \begin{bmatrix} -1 & 2 & 2 \\ -1 & 3 & 2 \\ 3 & -6 & -5 \end{bmatrix} \begin{bmatrix} 3 & 4 & 4 \\ -1 & 2 & 0 \\ 3 & 0 & 2 \end{bmatrix} = \begin{bmatrix} 1 & 0 & 0 \\ 0 & 2 & 0 \\ 0 & 0 & 2 \end{bmatrix}.$$

The result is verified.

Remark: In this case two eigenvalues are equal but we could get third eigenvector from the space of eigenvectors corresponding to equal eigenvalues i.e. 2. to produce three linearly independent vectors to write orthogonal matrix.

Definition 5.3: The $n \times n$ matrix A is orthogonal if and only if the columns (and rows) of A form a orthogonal set of vectors. If all the vectors are of unit magnitude, then matrix is called orthonormal.

Theorem 5.4: A $n \times n$ matrix is digonalizable if all the eigenvalues are real and distinct.

Further if all eigenvalues are not distinct but n linearly independent eigenvectors exist, then also A can be diagonalized.

If n linearly independent eigenvectors can not be formed in case of some equal eigenvalues then diagonalization of matrix A is not possible.

This fact is illustrated by the following examples.

Example 12: Consider the matrix $P = \begin{bmatrix} 3 & 4 & 3 \\ -4 & 5 & -3 \\ 1 & 3 & 2 \end{bmatrix}$.

Eigenvalues of A are $\lambda = 2, -1, -1$.

For eigenvalue $\lambda = 2$, eigenvector is $(1, -1, 1)^T$ and for $\lambda = -1$ eigenvector is $(1 -1, 0)^T$.

Now if third linearly independent vector is constructed $(0, 1, 1)^T$ (which is not an eigenvector).

Then orthogonal matrix $P = \begin{bmatrix} 1 & 1 & 0 \\ -1 & -1 & 1 \\ 1 & 0 & 1 \end{bmatrix}$ and it's inverse $P^{-1} = \begin{bmatrix} -1 & -1 & 1 \\ 2 & 1 & -1 \\ 1 & 1 & 0 \end{bmatrix}$.

Now $AP = \begin{bmatrix} 3 & 4 & 3 \\ -4 & 5 & -3 \\ 3 & 3 & 2 \end{bmatrix} \begin{bmatrix} 1 & 1 & 0 \\ -1 & -1 & 1 \\ 1 & 0 & 1 \end{bmatrix} = \begin{bmatrix} 2 & -1 & 7 \\ -12 & -9 & 2 \\ 2 & 0 & 5 \end{bmatrix}$

and $P^{-1}AP = \begin{bmatrix} -1 & -1 & 1 \\ 2 & 1 & -1 \\ 1 & 1 & 0 \end{bmatrix} \begin{bmatrix} 2 & -1 & 7 \\ -12 & -9 & 2 \\ 2 & 0 & 5 \end{bmatrix} = \begin{bmatrix} 12 & 10 & -4 \\ -10 & -11 & 11 \\ -10 & -10 & 9 \end{bmatrix}$

The resulted matrix on right side is not diagonal matrix.

5.4 DIAGONALIZATION OF SYMMETRIC MATRICES

Definition 5.4: A $n \times n$ matrix A is said to be symmetric if $A^T = A$, i.e., where A^T is transpose of A. In such a matrix diagonal entries may be any numbers but off diagonal entries $a_{ij} = a_{ji}$ $i \neq j$ are equal in pairs for $i, j = 1, 2, ..., n$.

Theorem 5.5: A $n \times n$ matrix A is diagonalizable. Proof of this theorem is omitted.

Some properties of eigenvalues and eigenvectors of a symmetric matrix $A_{n \times n}$ are:

(a) Number of real eigenvalues is n, counting multiplicities of an eigenvalue.

(b) Dimension of each eigenspace corresponding l_k is equal the multiplicities of λ_k and such eigenspaces are mutually orthogonal.

(c) A is orthogonally diagonalizable.

Definition 5.5: Two vector v_1 and v_2 are said to be orthogonal if $v_1^T v_2 = 0$, where $v_1 = (x_1, x_2, x_3)^T$, $v_2 = (y_1, y_2, y_3)^T$ i.e., $x_1 y_1 + x_2 y_2 + x_3 y_3 = 0$.

Theorem 5.6: If A is a symmetric matrix, then any two eigenvectors from different eigenspaces are orthogonal.

Proof: Let v_1 and v_2 be eigenvectors corresponding to distinct eigenvalues say λ_1 and λ_1.

Now to show that v_1 and v_2 are orthogonal i.e., $v_1^T. v_2 = 0$

$$\lambda_1(v_1^T.v_2) = (\lambda_1 v_1)^T.v_2 = (Av_1)^T.v_2$$

$$= v_1^T.A^T v_2 = v_1^T.Av_2, \text{ since } A^T = A, A \text{ is symmetric.}$$

$$= v_1^T.\lambda_2 v_2$$

$$= \lambda_2 v_1^T.v_2 = \lambda_2(v_1^T.v_2).$$

Now $(\lambda_1 - \lambda_2)(v_1^T.v_2) = 0$, since $\lambda_1 \neq \lambda_2$, therefore $(v_1^T.v_2) = 0$.

Hence v_1 and v_2 are orthogonal to each other.

Example 13: Consider symmetric matrix $A = \begin{bmatrix} -2 & 4 & -2 \\ 4 & 4 & -4 \\ -2 & -4 & 5 \end{bmatrix}$.

Eigenvalues of matrix A are $-4, 1, 10$ with corresponding eigenvectors $(2, -1, 0)^T$, $(2, 4, 5)^T$ and $(1, 2, -2)^T$.

Orthonormal matrix P constructed from these eigenvectors is

$$P = \begin{bmatrix} \dfrac{2}{\sqrt{5}} & \dfrac{2}{\sqrt{45}} & \dfrac{1}{3} \\ -\dfrac{1}{\sqrt{5}} & \dfrac{4}{\sqrt{45}} & \dfrac{2}{3} \\ 0 & \dfrac{5}{\sqrt{45}} & -\dfrac{2}{3} \end{bmatrix} \text{ and it's inverse matrix } P^{-1} = P^T = \begin{bmatrix} \dfrac{2}{\sqrt{5}} & -\dfrac{1}{\sqrt{5}} & 0 \\ \dfrac{2}{\sqrt{45}} & \dfrac{4}{\sqrt{45}} & \dfrac{5}{\sqrt{45}} \\ \dfrac{1}{3} & \dfrac{2}{3} & -\dfrac{1}{3} \end{bmatrix}.$$

Now $AP = \begin{bmatrix} -2 & 4 & -2 \\ 4 & 4 & -4 \\ -2 & -4 & 5 \end{bmatrix} \begin{bmatrix} \dfrac{2}{\sqrt{5}} & \dfrac{2}{\sqrt{45}} & \dfrac{1}{3} \\ -\dfrac{1}{\sqrt{5}} & \dfrac{4}{\sqrt{45}} & \dfrac{2}{3} \\ 0 & \dfrac{5}{\sqrt{45}} & -\dfrac{2}{3} \end{bmatrix} = \begin{bmatrix} -\dfrac{8}{\sqrt{5}} & \dfrac{2}{\sqrt{45}} & \dfrac{10}{3} \\ \dfrac{4}{\sqrt{5}} & \dfrac{4}{\sqrt{45}} & \dfrac{20}{3} \\ 0 & \dfrac{5}{\sqrt{45}} & -\dfrac{20}{3} \end{bmatrix},$

Eigenvalues and Eigenvectors

and $P^{-1} AP = \begin{bmatrix} \dfrac{2}{\sqrt{5}} & -\dfrac{1}{\sqrt{5}} & 0 \\ \dfrac{2}{\sqrt{45}} & \dfrac{4}{\sqrt{45}} & \dfrac{5}{\sqrt{45}} \\ \dfrac{1}{3} & \dfrac{2}{3} & -\dfrac{2}{3} \end{bmatrix} \begin{bmatrix} -\dfrac{8}{\sqrt{5}} & \dfrac{2}{\sqrt{45}} & \dfrac{10}{3} \\ \dfrac{4}{\sqrt{5}} & \dfrac{4}{\sqrt{45}} & \dfrac{20}{3} \\ 0 & \dfrac{5}{\sqrt{45}} & -\dfrac{20}{3} \end{bmatrix}$

$= \begin{bmatrix} -4 & 0 & 0 \\ 0 & 1 & 0 \\ 0 & 0 & 10 \end{bmatrix}$

This is the required diagonal form of the matrix.

Example 14: Consider the matrix $A = \begin{bmatrix} 3 & 2 & 4 \\ 2 & 0 & 2 \\ 4 & 2 & 3 \end{bmatrix}$, with eigenvalues $\lambda = 8, -1, -1$ and their corresponding eigenvectors (2, 1, 2). and [(1, −2, 0), (0, −2, 1)] − (0, 0, 0).

Vectors $\{(2, 1, 2), (1, -2, 0), (0, -2, 1)\}$ are not orthogonal, therefore changing vector (0, −2, 1) to (4, 2, −5), obtained by constructing orthogonal to (2, 1, 2) and (1, −2, 0).

Therefore to form orthogonal matrix, three orthogonal vectors are $\{(2, 1, 2), (1, -2, 0), (4, 2, -5)\}$, since (4, 2, −5) = 4(1, −2, 0) −5(0, −2, 1).

Now with the help of orthogonal vectors, orthonormal vectors constructed give

Orthonormal matrix $P = \begin{bmatrix} \dfrac{2}{3} & \dfrac{1}{\sqrt{5}} & \dfrac{4}{\sqrt{45}} \\ \dfrac{1}{3} & -\dfrac{2}{\sqrt{5}} & \dfrac{2}{\sqrt{45}} \\ \dfrac{2}{3} & 0 & -\dfrac{5}{\sqrt{45}} \end{bmatrix}$

and inverse matrix $P^{-1} = P^T = \begin{bmatrix} \dfrac{2}{3} & \dfrac{1}{3} & \dfrac{2}{3} \\ \dfrac{1}{\sqrt{5}} & -\dfrac{2}{\sqrt{5}} & 0 \\ \dfrac{4}{\sqrt{45}} & \dfrac{2}{\sqrt{45}} & -\dfrac{5}{\sqrt{45}} \end{bmatrix}$.

Now $AP = \begin{bmatrix} 3 & 2 & 4 \\ 2 & 0 & 2 \\ 4 & 2 & 3 \end{bmatrix} \begin{bmatrix} \frac{2}{3} & \frac{1}{\sqrt{5}} & \frac{4}{\sqrt{45}} \\ \frac{1}{3} & -\frac{2}{\sqrt{5}} & \frac{2}{\sqrt{45}} \\ \frac{2}{3} & 0 & -\frac{5}{\sqrt{45}} \end{bmatrix} = \begin{bmatrix} \frac{16}{3} & -\frac{1}{\sqrt{5}} & -\frac{4}{\sqrt{45}} \\ \frac{8}{3} & \frac{2}{\sqrt{5}} & -\frac{2}{\sqrt{45}} \\ \frac{16}{3} & 0 & \frac{5}{\sqrt{45}} \end{bmatrix}.$

and $P^{-1}AP = \begin{bmatrix} \frac{2}{3} & \frac{1}{3} & \frac{2}{3} \\ \frac{1}{\sqrt{5}} & -\frac{2}{\sqrt{5}} & 0 \\ \frac{4}{\sqrt{45}} & \frac{2}{\sqrt{45}} & -\frac{5}{\sqrt{45}} \end{bmatrix} \begin{bmatrix} \frac{16}{3} & -\frac{1}{\sqrt{5}} & -\frac{4}{\sqrt{45}} \\ \frac{8}{3} & \frac{2}{\sqrt{5}} & -\frac{2}{\sqrt{45}} \\ \frac{16}{3} & 0 & \frac{5}{\sqrt{45}} \end{bmatrix} = \begin{bmatrix} 8 & 0 & 0 \\ 0 & -1 & 0 \\ 0 & 0 & -1 \end{bmatrix}$

which is the required result.

5.5 QUADRATIC FORMS

Definition 5.6: A quadratic form on R^n is a function Q defined on R^n, whose value at a vector x in R^n can be computed by an expression of the form $Q(x) = x^T A x$, where A is an $n \times n$ symmetric matrix. The matrix A is called the matrix of the quadratic form.

For illustration let us consider vector $x = (x_1, x_2, x_3)^T$ in R^3 and a 3×3 matrix $A = (a_{ij})_{3 \times 3}$ with $a_{12} = a_{21}$, $a_{13} = a_{31}$, $a_{23} = a_{32}$.

For $x = (x_1, x_2, x_3)^T$ and $A = \begin{bmatrix} a_{11} & a_{12} & a_{13} \\ a_{21} & a_{22} & a_{23} \\ a_{31} & a_{32} & a_{33} \end{bmatrix}$,

$Q(x) = x^T A x = (x_1, x_2, x_3) \begin{bmatrix} a_{11} & a_{12} & a_{13} \\ a_{21} & a_{22} & a_{23} \\ a_{31} & a_{32} & a_{33} \end{bmatrix} \begin{pmatrix} x_1 \\ x_2 \\ x_3 \end{pmatrix}$

$= (x_1 a_{11} + x_2 a_{21} + x_3 a_{31}, x_1 a_{12} + x_2 a_{22} + x_3 a_{32}, x_1 a_{31} + x_2 a_{32} + x_3 a_{33}) \begin{pmatrix} x_1 \\ x_2 \\ x_3 \end{pmatrix}$

$$= a_{11}x_1^2 + a_{21}x_1x_2 + a_{31}x_3x_1 + a_{12}x_1x_2 + a_{22}x_2^2 + a_{32}x_2x_3 + a_{31}x_1x_3 + x_2a_{32}x_3 + a_{33}x_3^2$$
$$= a_{11}x_1^2 + a_{22}x_2^2 + a_{33}x_3^2 + 2a_{12}x_1x_2 + 2a_{23}x_2x_3 + 2a_{31}x_3x_1$$

since $a_{21} = a_{12}, a_{32} = a_{23}, a_{13} = a_{31}$.

This expression is quadratic form in R^3.

If $x^T A x$ = constant, then above equation represents ellipsoid, hyperboloid and so on.

In R^2, Quadratic form $(x, y)\begin{bmatrix} a & h \\ h & b \end{bmatrix}\begin{pmatrix} x \\ y \end{pmatrix} = c$ becomes $(ax + hy, xh + by)\begin{pmatrix} x \\ y \end{pmatrix} = c$

i.e. $ax^2 + 2hxy + by^2 = c$. A conic section centered at the origin in x-y plane two dimensional spaces.

While in xyz three dimension space, equation $ax^2 + by^2 + cz^2 + 2hxy + 2gzx + 2fyz = k$ is a quadratic surface centered at the origin where $a, b, c, h, g, f,$ and k are real numbers.

To know the nature of curve as ellipse, hyperbola etc., in two dimensions and the nature of surface as ellipsoid, hyperboloid, paraboloid etc., in three dimensions, rotation and translation of the coordinate axes is required. These can be understood with the help of eigenvalues and eigenvectors of the matrices associated with these quadratic forms.

As above, surface equation can be written in matrix form as

$$(x, y, z)\begin{bmatrix} a & h & f \\ h & b & g \\ f & g & c \end{bmatrix}\begin{pmatrix} x \\ y \\ z \end{pmatrix} = k,$$

Where $A = \begin{bmatrix} a & h & f \\ h & b & g \\ f & g & c \end{bmatrix}$ is matrix associated with the equation.

In short matrix form $x^T A x = k$, where $x^T = (x, y, z)$.

Theorem 5.7: Principal Axes Theorem: Let A be an $n \times n$ symmetric matrix, there is an orthogonal change of variable $x = Py$, that transforms the quadratic form $x^T Ax$ into quadratic form $y^T Dy$ with no cross product term. where D is diagonal.

Definition 5.7: A quadratic form $Q(x) = x^T Ax$ with an $n \times n$ symmetric matrix A, for x in R^n is known as

(a) Positive definite if $Q(x) > 0$ for all $x \neq 0$.
  Positive semi-definite if $Q(x) \geq 0$ for all x

(b) Negative definite if $Q(x) < 0$ for all $x \neq 0$
  Negative semi definite if $(Q(x) \leq 0$ for all x

(c) Indefinite if $Q(x)$ assumes both positive and negative values for different x.

5.22   Elementary Linear Algebra

Theorem 5.8: Let A be an $n \times n$ symmetric matrix. Then Quadratic form $x^T A x$ is
(a) Positive definite if and only if all the eigenvalues of A are positive.
(b) Negative definite if and only if all eigenvalues of A are negative.
(c) Indefinite if A has both positive and negative eigenvalues.

Example 15: Let $x = \begin{pmatrix} x_1 \\ x_2 \\ x_3 \end{pmatrix}$ and matrix $A = \begin{bmatrix} 3 & 2 & -1 \\ 2 & -5 & 3 \\ -1 & 3 & 4 \end{bmatrix}$

Compute $x^T A x$.

$$Ax = \begin{bmatrix} 3 & 2 & -1 \\ 2 & -5 & 3 \\ -1 & 3 & 4 \end{bmatrix} \begin{bmatrix} x_1 \\ x_2 \\ x_3 \end{bmatrix}$$

$$= \begin{bmatrix} 3x_1 + 2x_2 - x_3 \\ 2x_1 - 5x_2 + 3x_3 \\ -x_1 + 3x_2 + 4x_3 \end{bmatrix}$$

$$= 3x_1^2 + 2x_1 x_2 - x_3 x_1 + 2x_1 x_2 - 5x_2^2 + 3x_2 x_3 - x_1 x_3 + 3x_2 x_3 + 4x_3^2$$

$$= 3x_1^2 - 5x_2^2 + 4x_3^2 + 4x_1 x_2 + 6x_2 x_3 - 2x_1 x_3.$$

Example 16: Find the matrix of the quadratic form

$$4x_1^2 + 3x_2^2 - 5x_3^2 + 6x_1 x_2 - 4x_2 x_3 + 8x_3 x_1$$

Arrange the coefficients of square terms along the diagonal and $\dfrac{1}{2}$ of coefficient of $x_1 x_2$ at first row and second column, and so on, then required matrix

$$A = \begin{bmatrix} 4 & 3 & 4 \\ 3 & 3 & -2 \\ 4 & -2 & -5 \end{bmatrix}$$

since $\quad x^T A x = (x_1, x_2, x_3) \begin{bmatrix} 4 & 3 & 4 \\ 3 & 3 & -2 \\ 4 & -2 & -5 \end{bmatrix} \begin{pmatrix} x_1 \\ x_2 \\ x_3 \end{pmatrix}$

Geometrically, we noted that in R^2, general quadratic form $ax^2 + 2hxy + by^2 = c$ represents a conic section with centre at $(0, 0)$. If the principal axes are along x-axes any

y axis, then cross product term xy should not appear. It can be done by rotation of axes on replacing $\begin{bmatrix} x \\ y \end{bmatrix} = \begin{bmatrix} a_{11} & a_{12} \\ a_{12} & a_{22} \end{bmatrix} \begin{bmatrix} x^1 \\ y^1 \end{bmatrix}$, such that $ax^2 + 2hxy + by^2 = c$ changes to $a^1 x^{1^2} + b^1 y^{1^2} = c$.

If $P = \begin{bmatrix} a_{11} & a_{12} \\ a_{12} & a_{22} \end{bmatrix}$, $\begin{pmatrix} x \\ y \end{pmatrix} = x$, $\begin{pmatrix} x^1 \\ y^1 \end{pmatrix} = y$

$x = Py$ or $y = P^{-1}x$ where P is invetible matrix

$$x^T Ax = (Py)^T APy = y^T P^T APy = y^T Dy$$

if P is orthogonal and diagonalizes A.

then $\qquad P^T = P^{-1}$

i.e. $\qquad P^T AP = P^{-1} AP = D$.

Therefore such a matrix P should be constructed from the orthogonal eigenvectors of A.

Example 17: Reduce $5x^2 + 6xy + 5y^2 = 4$ to the principal axes form.

$$(x, y) \begin{bmatrix} 5 & 3 \\ 3 & 5 \end{bmatrix} \begin{pmatrix} x \\ y \end{pmatrix} = 4.$$

On placing coefficient of quadratic terms along diagonal and $\dfrac{1}{2}$ of coefficient of cross product term xy in off diagonal places.

Matrix associated with quadratic form is $A = \begin{bmatrix} 5 & 3 \\ 3 & 5 \end{bmatrix}$, eigenvalues of A are 2, 8 and eigenvectors $\begin{pmatrix} 1 \\ -1 \end{pmatrix}, \begin{pmatrix} 1 \\ 1 \end{pmatrix}$.

Therefore orthonormal matrix $P = \begin{bmatrix} \dfrac{1}{\sqrt{2}} & \dfrac{1}{\sqrt{2}} \\ -\dfrac{1}{\sqrt{2}} & \dfrac{1}{\sqrt{2}} \end{bmatrix}$ and $P^{-1} = P^T = \begin{bmatrix} \dfrac{1}{\sqrt{2}} & -\dfrac{1}{\sqrt{2}} \\ \dfrac{1}{\sqrt{2}} & \dfrac{1}{\sqrt{2}} \end{bmatrix}$

with
$$y = p^{-1}x = \begin{bmatrix} \dfrac{1}{\sqrt{2}} & -\dfrac{1}{\sqrt{2}} \\ \dfrac{1}{\sqrt{2}} & \dfrac{1}{\sqrt{2}} \end{bmatrix} \begin{pmatrix} x \\ y \end{pmatrix};$$

therefore
$$\begin{bmatrix} x^1 \\ y^1 \end{bmatrix} = \begin{bmatrix} \dfrac{x-y}{\sqrt{2}} \\ \dfrac{x+y}{\sqrt{2}} \end{bmatrix}$$

$$5x^2 + 6xy + 5y^2 = 0; \text{ changes to}$$

$$x^T Ax = [x', y'] \begin{bmatrix} 2 & 0 \\ 0 & 8 \end{bmatrix} \begin{bmatrix} x' \\ y' \end{bmatrix} = 0; \; 2x'^2 + 8y'^2 = 4.$$

Example 18: Reduce the quadratic surface

$$-2x^2 + 4y^2 + 5z^2 + 8xy - 4xz - 8yz = 40$$

to the principal axes form.

Quadratic form is

$$(x, y, z) \begin{bmatrix} -2 & 4 & -2 \\ 4 & 4 & -4 \\ -2 & -4 & 5 \end{bmatrix} \begin{bmatrix} x \\ y \\ z \end{bmatrix} = 40$$

Matrix associated with above form is

$$A = \begin{bmatrix} -2 & 4 & -2 \\ 4 & 4 & -4 \\ -2 & -4 & 5 \end{bmatrix}, \text{ with eigenvalues } -4, 1, 10;$$

and corresponding orthonormal matrix $P^{-1} = \begin{bmatrix} \dfrac{2}{\sqrt{5}} & -\dfrac{1}{\sqrt{5}} & 0 \\ \dfrac{2}{\sqrt{45}} & \dfrac{4}{\sqrt{45}} & \dfrac{5}{\sqrt{45}} \\ \dfrac{1}{3} & \dfrac{2}{3} & -\dfrac{2}{3} \end{bmatrix}$

Principal form is $(x', y', z') \begin{bmatrix} -4 & 0 & 0 \\ 0 & 1 & 0 \\ 0 & 0 & 10 \end{bmatrix} \begin{bmatrix} x' \\ y' \\ z' \end{bmatrix} = 40$

$-4x'^2 + y'^2 + 10z'^2 = 40$ where $x = Py$ is

$$\begin{bmatrix} x \\ y \\ z \end{bmatrix} = \begin{bmatrix} \frac{2}{\sqrt{5}} & \frac{2}{\sqrt{45}} & \frac{1}{3} \\ -\frac{2}{\sqrt{5}} & \frac{4}{\sqrt{45}} & \frac{2}{3} \\ 0 & \frac{5}{\sqrt{45}} & -\frac{2}{3} \end{bmatrix} \begin{bmatrix} x' \\ y' \\ z' \end{bmatrix}$$

Quadratic surface is elliptic hyperboloid.

EXERCISE SET 5

1. Find all the eigenvalues and corresponding eigenvectors of the matrix
$$A = \begin{bmatrix} 2 & 4 \\ 5 & 3 \end{bmatrix}.$$

2. Find all the eigenvalues and corresponding eigenvectors of the matrix
$$A = \begin{bmatrix} 5 & 4 \\ 5 & 6 \end{bmatrix}.$$

3. Find all the eigenvalues and corresponding eigenvectors of the matrix
$$A = \begin{bmatrix} 2 & 1 & 1 \\ 1 & 2 & 1 \\ 1 & 1 & 2 \end{bmatrix}.$$

Also write a matrix with same eigenvalues.

4. Find all the eigenvalues and corresponding eigenvectors of the matrix
$$A = \begin{bmatrix} 3 & -1 & 2 \\ -1 & 2 & 1 \\ 2 & 1 & 3 \end{bmatrix}$$

Write another 3×3 matrix, which has same eigenvalues as matrix A.

5. Find all the eigenvalues and corresponding eigenvectors of the matrix

$$A = \begin{bmatrix} 8 & -6 & 2 \\ -6 & 7 & -4 \\ 2 & -4 & 3 \end{bmatrix}$$

6. Obtain the eigenvalues and eigenvectors of

$$A = \begin{bmatrix} -2 & 2 & 1 \\ 2 & 1 & 2 \\ 1 & 2 & 6 \end{bmatrix}$$

and prove vectors are pairwise orthogonal.

7. Let matrix $\begin{bmatrix} -1 & 2 & 1 \\ -4 & 5 & 1 \\ -1 & -2 & -3 \end{bmatrix}$ and matrix $B = 3A^2 - 4A + I$, where I is 3×3 identity matrix. Find the eigenvalues and corresponding eigenvectors of the matrix B.

8. Given the eigenvalues $\lambda = -4, 0, 3$ and corresponding eigenvectors
$[(1, 2, -1)]^T - (0, 0, 0), [(1, 6, -1\ 3)]^T - (0, 0, 0)$, and $[(2, 3, -2)]^T - (0, 0, 0)$

of the matrix $A = \begin{bmatrix} 1 & 2 & 1 \\ 6 & -1 & 0 \\ -1 & -2 & -1 \end{bmatrix}$, find the eigenvalues and corresponding

eigenvectors of the matrix B, where $B = 3A^2 + 4A - 5I$.

9. Compute the eigenvalues and corresponding eigenvectors of A^{-1}, where matrix

$A = \begin{bmatrix} 4 & 2 & 5 \\ 3 & 3 & 5 \\ 3 & 2 & 6 \end{bmatrix}$. Hence compute the eigenvalues and corresponding eigenvectors of

$4A^{-2} - 3A^{-1} + 2I$.

10. Let matrix $A = \begin{bmatrix} 3 & -1 & 2 \\ -1 & 2 & 1 \\ 2 & -1 & 3 \end{bmatrix}$. Find the eigenvalues and corresponding

eigenvectors of the matrix B, where $B = 2(A)^2 + 3A - 5I$ where, I is 3×3 identity matrix.

Eigenvalues and Eigenvectors

11. Diagonalize the following given matrix A by an orthonormal transformation:

$$A = \begin{bmatrix} 5 & -6 & -6 \\ -1 & 4 & 2 \\ 3 & -6 & -4 \end{bmatrix}.$$

12. Diagonalize the following given matrix A by an orthonormal transformation:

$$A = \begin{bmatrix} 3 & -1 & 2 \\ -1 & 2 & 1 \\ 2 & -1 & 3 \end{bmatrix}.$$

13. Using orthonormal matrices, diagonalize the symmetric matrix

$$A = \begin{bmatrix} 4 & 2 & 0 \\ 2 & 3 & -2 \\ 0 & -2 & 2 \end{bmatrix}.$$

14. Find all the eigenvalues and corresponding eigenvectors of the matrix A:

$$A = \begin{bmatrix} 2 & 1 & 1 \\ 1 & 2 & 1 \\ 1 & 1 & 4 \end{bmatrix}.$$

   Hence diagonalize the matrix A by using orthonormal matrix.

15. Find the matrix of the quadratic form $2x^2 + y^2 + 2z^2 - 2xy + 4xz + 2yz + 16$ and hence reduce the above form into principal axes form.

16. Find the matrix of the quadratic form $-x^2 + 2y^2 + 7z^2 + 4xy + 2xz + 4yz - 9$ and hence reduce the above form into principal axes form.

17. Find the quadratic form for the matrix A given by

$$A = \begin{bmatrix} 4 & 2 & 0 \\ 2 & 3 & -2 \\ 0 & -2 & 2 \end{bmatrix}.$$

18. Find the quadratic form for the matrix A given by

$$A = \begin{bmatrix} 7 & 2 & 3 \\ 2 & 2 & -3 \\ 3 & -3 & 6 \end{bmatrix}.$$

ANSWERS TO EXERCISE – 5

1. Eigen values $\lambda = -2, 7$ Eigen vectors $\alpha (1, -1)$, $\alpha \neq 0$, $[\beta (4, 5)]$ $\beta \neq 0$,
2. Eigen values $\lambda = 1, 10$
   Eigen vectors $\alpha (1, -1)$, $\alpha \neq 0$, $\beta(4, 5)$, $\beta \neq 0$,
3. $\lambda = 1, 1, 4$ eigen vectors $[(1, 0, -1), (0, 1, -1) \setminus (0, 0, 0)]$ $\alpha (1, 1, 1)$, and $\alpha \neq 0$

   Matrix with same eigenvalues $\begin{bmatrix} 3 & 2 & 2 \\ 2 & 3 & 2 \\ 2 & 2 & 3 \end{bmatrix}$.

4. $\lambda = 0, 3, 5$; eigen vector $X_1 = \alpha (1, 1, -1)^T$, $\alpha \neq 0$, $X_2 = \beta(-1, 2, 1)^T$, $\beta \neq 0$, $X_3 = \gamma(1, 0, 1)^T$, $\gamma \neq 0$
5. $\lambda = 0, 3, 15$; eigen vectors $X_1 = (1, 2, 2)^T$, $X_2 = (2, 1, -2)^T$, $X_3 = (2, -2, 1)^T$ are orthogonal

6. $\lambda = 1, -3, 7$; eigenvectors $x_1 = \begin{bmatrix} 1 \\ 2 \\ -1 \end{bmatrix}, x_2 = \begin{bmatrix} -2 \\ 1 \\ 0 \end{bmatrix}, x_3 = \begin{bmatrix} 1 \\ 2 \\ 5 \end{bmatrix}$.

7. Eigen values of A are $\lambda = -2, 0, 3$; eigenvectors $X_1 = (1, 1, -3)^T$, $X_2 = (1, 1, -1)^T$, $X_3 = (2, 5, -2)^T$, Eigen values of B are $\lambda = 21, 1, 16$; eigenvectors will remain the same
8. Eigen values of B are $\lambda = 29, -5, 34$; eigenvectors will remain the same
9. Eigenvalues of A are $\lambda = 1, 1, 11$, eigenvector $[(2, -3, 0), (0, -5, 2)] \setminus (0, 0, 0)$ for $\lambda = 1$, $\alpha (1, 1, 1)$, $\alpha \neq 0$ for $\lambda = 11$, eigenvalues of B are $\lambda = 2, 2, 3$; eigenvectors are same.
10. Eigen values of A, $\lambda = 1, 2, 5$ and eigenvalues of B are $-5, 22, 60$, eigenvectors are same as of A.

11. $\begin{bmatrix} 5 & 0 & 0 \\ 0 & 3 & 0 \\ 0 & 0 & 0 \end{bmatrix}$
12. $\begin{bmatrix} 1 & 0 & 0 \\ 0 & 2 & 0 \\ 0 & 0 & 5 \end{bmatrix}$

13. Eigenvalues $\lambda = 6, 3, 0$, diagonalized matrix $\begin{bmatrix} 6 & 0 & 0 \\ 0 & 3 & 0 \\ 0 & 0 & 0 \end{bmatrix}$

14. Eigenvalues of A are $\lambda = 1, 2, 5$, eigenvector $\alpha\,(1, -1, 0)$, $\beta\,(1, 1, 0)]$, $\lambda(1, 1, 2)$ $\alpha, \beta, \gamma \neq 0$; diagonalized matrix

$$\begin{bmatrix} 1 & 0 & 0 \\ 0 & 2 & 0 \\ 0 & 0 & 5 \end{bmatrix}.$$

15. Matrix $\begin{bmatrix} 2 & -1 & 2 \\ -1 & 1 & 1 \\ 2 & 1 & 2 \end{bmatrix}$, principal axis form $x^2 - 2y^2 - 4z^2 = 16$.

16. Matrix $\begin{bmatrix} -1 & 2 & 1 \\ 2 & 2 & 2 \\ 1 & 2 & 7 \end{bmatrix}$ principal axis form $2x^2 - 2y^2 + 8z^2 = 9$.

17. Quadratic form $4x_1^2 + 3x_2^2 + 2x_3^2 + 4x_1 x_2 - 4x_2 x_3$.

18. Quadratic form $7x_1^2 + 2x_2^2 + 6x_3^2 + 4x_1 x_2 + 6x_1 x_3 - 6x_2 x_3$.

CHAPTER 6

Inner Product

As we have noticed that two dimensional x-y space and three-dimensional xyz space are examples of vector spaces with usual properties. In geometry, we have concept of length and orthogonality also. Therefore these concepts are also extended to general vector spaces with the help of inner product of any two vectors of that space.

6.1 INNER PRODUCT

Definition 6.1: Let V be a vector space in general (finite or infinite dimensional) with real scalars since in this text, only real scalars have been taken for vector spaces. Inner product of two vectors is a rule which assigns a real number denoted by $\langle u, v \rangle$ to each pair of vectors $u, v \in V$ such that

(a) $\langle u, v \rangle = \langle v, u \rangle$

(b) $\langle (u+v), w \rangle = \langle u, w \rangle + \langle v, w \rangle$, for all $u, v, w \in V$

(c) $\langle \alpha u, v \rangle = \alpha \langle u, v \rangle = \langle u, \alpha v \rangle$

(d) $\langle u, u \rangle \geq 0$, and $\langle u, u \rangle = 0$ if and only if $u = 0$.

Note in case of complex scalars condition (c) is $\langle \alpha u, v \rangle = \langle u, \overline{\alpha} v \rangle$, where $\overline{\alpha}$ is complex conjugate of α.

Then the length of vector of u is defined as a positive real number $\sqrt{\langle u, u \rangle}$ and denoted by $\|\alpha\|$ and is read as norm of u.

Example 1: Let $V = R^n$ be space of n-tuples i.e.

$$V = \{u = (x_1, x_2, ..., x_n) \mid x_i's \text{ are real numbers}\} = R^n$$

Further Let $v = (y_1, y_2, ..., y_n) \in R^n$.

$\langle u, v \rangle = x_1 y_1 + x_2 y_2 + , ..., + x_n y_n$ is defined as inner product in R^n.

Conditions (a), (b), and (c) are obvious. Now condition (d) is being checked

(d) $\langle u, u \rangle = x_1^2 + ... + x_n^2$ is sum of squares of real numbers therefore $\langle u, u \rangle \geq 0$, if $\langle u, u \rangle = x_1^2 + ... + x_n^2 = 0 \Leftrightarrow x_1 = x_2 = ... = x_n = 0 \Leftrightarrow u = (0, 0, ..., 0)$, i.e. zero 0_V vector of R^n.

Note: $\sqrt{\langle u, u \rangle} = \|u\| = \sqrt{x_1^2 + x_2^2 + ... + x_n^2}$ agrees with usual formula of length in geometry in R^n. Scalar product of vectors in two dimensional x-y plane or three dimensional xyz-space is inner pruduct.

Example 2: Let $V = C[a, b]$ be the space of real valued continuous functions on the interval $[a, b]$.

Inner product defined for $f, g \in C[a, b]$ is as $\langle f, g \rangle = \int_a^b f(x) g(x) dx$.

Again conditions (a) (b) (c) are obvious.

To check (d) $\langle f, f \rangle = \int_a^b [f(x)]^2 dx$ integral of a non-negative function on $[a, b]$ is always non-negative. And if $\langle f, f \rangle = 0 \Rightarrow \int_a^b [f(x)]^2 dx = 0 \Leftrightarrow f(x) \equiv 0$ on $[a, b]$, hence above defined rule is an inner product on $C[a, b]$.

In general $w(x) \geq 0$ on $[a, b]$ is a weight function, then inner product defined in $C[a, b]$ is $\langle f, g \rangle = \int_a^b w(x) f(x) g(x) dx$ for $f, g \in C[a, b]$.

Example 3: Inner product can be defined on a vector space in more than one ways as shown in this example.

Let $V = P_n$ be the space of polynomials of degree $\leq n$.

Let $p(x), q(x) \in P_n$, define $\langle p(x), q(x) \rangle = \int_a^b p(x) q(x) dx$ since polynomials are continuous on R, therefore from above example, this is an inner product. Another way of defining inner product on polynomial space is as, let $p(x) = a_0 + a_1 x + , ..., + a_n x^n$ and $q(x) = b_0 + b_1 x + , ..., + b_n x^n$, then $<p(x), q(x)> = a_0 b_0 + a_1 b_1 + ... + a_n b_n$.

Again conditions (a), (b) and (c) are obviously satisfied. To check condition (d)
$\langle p(x), p(x) \rangle = a_0^2 + a_1^2 + ,..., + a_n^2 = 0 \Leftrightarrow a_0 = a_1 = a_n = 0 \Leftrightarrow p(x) = 0.$

Hence above rule defines inner product on space of polynomials P_n.

Length of $p(x) = \|p(x)\| = \sqrt{a_0^2 + a_1^2 + ,..., + a_n^2}$.

6.2 ORTHOGONALITY

Definition 6.2 Let V be a vector space on which an inner product is defined. Two vectors $u, v \in V$ are called orthogonal if $\langle u, v \rangle = 0$.

A set $S = \{u_1, u_2, ..., u_n\}$ of vectors of V is orthogonal if pair wise all vectors of S are orthogonal i.e $\langle u_i, u_j \rangle = 0$. for all $i, j = 1, 2, ..., n, i \neq j$.

Further an infinite set S is orthogonal if every finite subset of S is orthogonal in itself.

Remark: Zero vector 0_V is orthogonal to each vector in the vector space V.

Orthonormal Set: A set $S = \{u_1, u_2, u_3, ..., u_n\}$ of a vector space V is known as orthonormal, if it is orthogonal and each vector of S is of unit length i.e., unit magnitude.

Example 4: Let for any and $u = (x_1, x_2, x_3), v = (y_1, y_2, y_3) \in R^3$ inner product be defined by $\langle u, v \rangle = x_1 y_1 + x_2 y_2 + x_3 y_3$, then Set $\{(1, 0, 0), (0, 1, 0), (0, 0, 1)\}$ is orthonormal in R^3,

$\langle (1, 0, 0), (0, 1, 0) \rangle = 1 \times 0 + 0 \times 1 + 0 \times 0 = 0,$

$\langle (0, 1, 0), (0, 0, 1) \rangle = 0 \times 0 + 1 \times 0 + 0 \times 1 = 0,$

$\langle (0, 0, 1), (1, 0, 0) \rangle = 0 \times 1 + 0 \times 0 + 1 \times 0 = 0.$

Pair wise all three vectors are orthogonal, therefore given set is orthogonal. Now to prove orthonormal,

$\langle (1, 0, 0), (1, 0, 0) \rangle = 1 \times 1 + 0 \times 0 + 0 \times 0 = 1 \Rightarrow \|(1, 0, 0)\| = 1,$

$\langle (0, 1, 0), (0, 1, 0) \rangle = 0 \times 0 + 1 \times 1 + 0 \times 0 = 1 \Rightarrow \|(0, 1, 0)\| = 1,$

$\langle (0, 0, 1), (0, 0, 1) \rangle = 0 \times 0 + 0 \times 0 + 1 \times 1 = 1 \Rightarrow \|(0, 0, 1)\| = 1.$

Length of each vector is one, hence set is Orthonormal.

Example 5: Set $S = \{(1, -1, 0, 1), (0, 1, -1, 1), (-1, 0, 1, 1), (1, 1, 1, 0)\}$ is orthogonal with usual inner product in R^4, but not orthonormal because magnitude of each vector is $\sqrt{3}$.

$\langle(1,-1,0,1),(0,1,-1,1)\rangle = 1 \times 0 - 1 \times 1 + 0 \times -1 + 1 \times 1 = -1 + 1 = 0$

$\langle(1,-1,0,1),(-1,0,1,1)\rangle = 1 \times -1 - 1 \times 0 + 0 \times 1 + 1 \times 1 = 0$

$\langle(1,-1,0,1),(1,1,1,0)\rangle = 1 \times 1 - 1 \times 1 + 0 \times 1 + 1 \times 0 = 0$

$\langle(0,1,-1,1),(-1,0,1,1)\rangle = 0 \times -1 + 1 \times 0 - 1 \times 1 + 1 \times 1 = -1 + 1 = 0$

$\langle(0,1,-1,1),(1,1,1,0)\rangle = 0 \times 1 + 1 \times 1 - 1 \times 1 + 1 \times 0 = 1 - 1 = 0$

$\langle(-1,0,1,1),(1,1,1,0)\rangle = -1 \times 1 + 0 \times 1 + 1 \times 1 + 1 \times 0 = -1 + 1 = 0.$

Pair wise all four vectors are orthogonal. Therefore given set is orthogonal.

Now magnitude of each vector = $\sqrt{1+1+1} = \sqrt{3}$. Therefore given set is not Orthonormal.

Now unit vectors are obtained by dividing each vector by $\sqrt{3}$. Therefore orthonormal set is

$$S_1 = \left\{\left(\frac{1}{\sqrt{3}}, \frac{-1}{\sqrt{3}}, 0, \frac{1}{\sqrt{3}}\right), \left(0, \frac{1}{\sqrt{3}}, \frac{-1}{\sqrt{3}}, \frac{1}{\sqrt{3}}\right), \left(-\frac{1}{\sqrt{3}}, 0, \frac{1}{\sqrt{3}}, \frac{1}{\sqrt{3}}\right), \left(\frac{1}{\sqrt{3}}, \frac{1}{\sqrt{3}}, \frac{1}{\sqrt{3}}, 0\right)\right\}.$$

Theorem 6.1: If $S = \{u_1, u_2, ..., u_n\}$ is an orthogonal set of non-zero vectors in an n-dimensional vector space V, then S is linearly independent.

Proof: Consider linear combination

$$\alpha_1 u_1 + \alpha_2 u_2 + , ..., + \alpha_n u_n = 0_V.$$

On taking inner product with u_i on both sides, we get

$$\langle u_i, (\alpha_1 u_1 + \alpha_2 u_2 + , ..., + \alpha_n u_n)\rangle = \langle u_i, 0_V\rangle = 0 \text{ for } i = 1, 2, ..., n.$$

Equivalently, we can write

$$\alpha_1 \langle u_i, u_1\rangle + \alpha_2 \langle u_i, u_2\rangle + \alpha_3 \langle u_i, u_3\rangle + ... + \langle u_i, u_i\rangle + ... + \alpha_n \langle u_i, u_n\rangle = 0$$

Since S is orthogonal $\langle u_i, u_j\rangle = 0$ for $j \neq i$. $i, j = 1, 2, ...n.$,

$\alpha_i \langle u_i, u_i\rangle = 0$ but $\langle u_i, u_i\rangle \neq 0$, since $u_i \neq 0_V$, therefore $\alpha_i = 0$, $i = 1, 2, ... n.$

Therefore all scalars, $\alpha_i = 0$, $i = 0, 1, 2, ..., n$.

Hence $S = \{u_1, u_2, ..., u_n\}$ is linearly independent.

Definition 6.3: A basis of a vector space V is called orthogonal basis, if all its vectors are orthogonal.

ORTHOGONAL PROJECTION

Given a non-zero vector **u** in an n-dimensional vector space V, let **v** be any other vector in V. Consider $v = \alpha u + w$ for some scalar α and vector w in V. We choose α and w such that $w = v - \alpha u$ is orthogonal to u i.e. $\langle w, u \rangle = \langle (v - \alpha u), u \rangle = \langle v, u \rangle - \alpha \langle u, u \rangle = 0$.

On solving above, we get $\alpha = \dfrac{\langle v, u \rangle}{\langle u, u \rangle}$.

Thus $w = v - \dfrac{\langle v, u \rangle}{\langle u, u \rangle} u$ and $\alpha u = \dfrac{\langle v, u \rangle}{\langle u, u \rangle} u$ is called orthogonal projection of v on u and $\dfrac{\langle v, u \rangle}{\langle u, u \rangle} u + w$ is orthogonal decomposition of v, where $w = v - \dfrac{\langle v, u \rangle}{\langle u, u \rangle} u$.

Set $S = \{1, x, x^2, x^3, x^4\}$ of P_4 is linearly independent, which is not orthogonal. For various applications a set of orthogonal set of vectors is required. The procedure of generating orthogonal vectors, from a given set of linearly independent vectors, is shown in the next sections.

6.3 GRAM-SCHMIDT ORTHOGONALIZATION

Given a set of linearly independent vectors $S = \{u_1, u_2, ..., u_n\}$ in a vector space V, generating the set of orthogonal or orthonormal vectors is known as Gram-Schmidt Orthogonalization Process.

The detailed procedure of Orthogonalization is as follows: begin with $u_1 \neq 0_V$ vector, because no vector in S is zero vector, since S is linearly independent.

So set $v_1 = u_1$.

Now to find next orthogonal vector (say) v_2.

Set $v_2 = u_2 + \alpha_1 v_1$, and choose α_1 such that $\langle v_2, v_1 \rangle = 0$.

i.e., $\langle v_2, v_1 \rangle = \langle u_2, v_1 \rangle + \alpha_1 \langle v_1, v_1 \rangle = 0 \Rightarrow \alpha_1 = -\dfrac{\langle u_2, v_1 \rangle}{\langle v_1, v_1 \rangle}$.

Thus $v_2 = u_2 - \dfrac{\langle u_2, v_1 \rangle}{\langle v_1, v_1 \rangle} v_1$ is orthogonal to v_1.

To get next vector v_3, set $v_3 = u_3 + \beta_1 v_1 + \beta_2 v_2$, since v_3 has to be orthogonal to v_1 and v_2, therefore

6.6 Elementary Linear Algebra

$$0 = \langle v_3, v_1 \rangle = \langle u_3, v_1 \rangle + \beta_1 \langle v_1, v_1 \rangle + \beta_2 \langle v_2, v_1 \rangle$$
$$= \langle u_3, v_1 \rangle + \beta_1 \langle v_1, v_1 \rangle, \text{ since } \langle v_2, v_1 \rangle = 0$$

therefore $\quad \beta_1 = -\dfrac{\langle u_3, v_1 \rangle}{\langle v_1, v_1 \rangle}$,

and $\quad 0 = \langle v_3, v_2 \rangle = \langle u_3, v_2 \rangle + \beta_1 \langle v_1, v_2 \rangle + \beta_2 \langle v_2, v_2 \rangle$
$$= \langle u_3, v_2 \rangle + \beta_2 \langle v_2, v_2 \rangle, \text{ since } \langle v_1, v_2 \rangle = 0$$

therefore $\quad \beta_2 = -\dfrac{\langle u_3, v_2 \rangle}{\langle v_2, v_2 \rangle}$

Thus $\quad v_3 = u_3 - \dfrac{\langle u_3, v_1 \rangle}{\langle v_1, v_1 \rangle} v_1 - \dfrac{\langle u_3, v_2 \rangle}{\langle v_2, v_2 \rangle} v_2.$

Continuing similarly, we can generate a set of orthogonal vectors, then we can get orthonormal set, dividing by their magnitude.

Example 6: Given $\{1, x, x^2, ..., x^n\}$ a linearly independent set in P_n, generate a sequence of orthogonal polynomials with inner product, $\langle p(x), g(x) \rangle = \int_{-1}^{1} p(x)g(x)dx$.

Let $p_0(x) = 1$ be a non-zero starting polynomial of degree zero.

Let $p_1(x) = x + \alpha$, where $\alpha = -\dfrac{\langle x, 1 \rangle}{\langle 1, 1 \rangle} = 0$ since $\langle x, 1 \rangle = \int_{-1}^{1} x \, dx = 0$, $\langle 1, 1 \rangle = \int_{-1}^{1} dx = 2$.

Thus $p_1(x) = x$, a first degree polynomial.
For second degree polynomial, let

$$p_2(x) = x^2 + \beta_1 1 + \beta_2 x, \text{ where } \beta_1 = -\dfrac{\langle x^2, 1 \rangle}{\langle 1, 1 \rangle}, \beta_2 = -\dfrac{\langle x^2, x \rangle}{\langle x, x \rangle}.$$

Now $\langle 1, x^2 \rangle = \int_{-1}^{1} x^2 dx = \dfrac{2}{3}$, $\langle x, x^2 \rangle = \int_{-1}^{1} x^3 dx = 0$, $<x, x> = \dfrac{2}{3}$.

Thus $p_2(x) = x^2 - \dfrac{2}{3}$.

For third degree polynomial, let $p_3(x) = x^3 + \gamma_1 1 + \gamma_2 x + \gamma_3 \left(x^2 - \dfrac{2}{3} \right).$

Inner Product

With $\gamma_1 = -\dfrac{\langle x^3, 1\rangle}{\langle 1,1\rangle}$, $\gamma_2 = -\dfrac{\langle x^3, x\rangle}{\langle x, x\rangle}$, $\gamma_3 = \dfrac{\left\langle x^3, \left(x^2 - \dfrac{2}{3}\right)\right\rangle}{\left\langle \left(x^2 - \dfrac{2}{3}\right), \left(x^2 - \dfrac{2}{3}\right)\right\rangle}$.

Now $\langle x, x\rangle = \displaystyle\int_{-1}^{1} x^2\, dx = \dfrac{2}{3}$, $\langle x^3, 1\rangle = \displaystyle\int_{-1}^{1} x^3\, dx = 0$, $\langle x^3, x\rangle = \displaystyle\int_{-1}^{1} x^4\, dx = \dfrac{2}{5}$, and

$\left\langle x^3, \left(x^2 - \dfrac{2}{3}\right)\right\rangle = \displaystyle\int_{-1}^{1} x^3\left(x^2 - \dfrac{2}{3}\right) dx = \displaystyle\int_{-1}^{1}\left(x^5 - \dfrac{2}{3}x^3\right) dx = 0$,

$\left\langle \left(x^2 - \dfrac{2}{3}\right), \left(x^2 - \dfrac{2}{3}\right)\right\rangle = \displaystyle\int_{-1}^{1}\left(x^2 - \dfrac{2}{3}\right)^2 dx \neq 0$,

Therefore $\gamma_1 = 0$, $\gamma_2 = -\dfrac{3}{5}$, $\gamma_3 = 0$.

Thus $p_3(x) = x^3 - \dfrac{3}{5}x$.

Similarly procedure can be continued to get next orthogonal polynomials.

Further, to get orthonormal set of polynomials, orthogonal polynomial vectors, obtained above, can be normalized by dividing by their magnitude.

As in this example $p_0(x) = 1$, $p_1(x) = x$, $p_2(x) = x^2 - \dfrac{2}{3}$ and $p_3(x) = x^3 - \dfrac{3}{3}x$.

Polynomials are to be divided by their magnitude to make orthonormal.

On computing inner product

$\langle 1, 1\rangle = 2$, $\langle x, x\rangle = \displaystyle\int_{-1}^{1} x^2\, dx = \dfrac{2}{3}$,

$\left\langle x^2 - \dfrac{2}{3}, x^2 - \dfrac{2}{3}\right\rangle = \displaystyle\int_{-1}^{1}\left(x^4 - \dfrac{4}{3}x^2 + \dfrac{4}{9}\right) dx = \dfrac{2}{5}$,

$\left\langle x^3 - \dfrac{3}{5}x, x^3 - \dfrac{3}{5}x\right\rangle = \displaystyle\int_{-1}^{1}\left(x^6 - \dfrac{6}{5}x^4 + \dfrac{9}{25}x^2\right) dx = \dfrac{8}{175}$.

We get $\|p_0(x)\| = \sqrt{\langle 1,1\rangle} = \sqrt{2}$, $\|p_1(x)\| = \sqrt{\langle x, x\rangle} = \sqrt{\dfrac{2}{3}}$,

$$\|p_2(x)\| = \sqrt{\left\langle \left(x^2 - \frac{2}{3}\right), \left(x^2 - \frac{2}{3}\right)\right\rangle} = \sqrt{\frac{2}{5}}$$

$$\|p_3(x)\| = \sqrt{\left\langle \left(x^3 - \frac{3}{5}x\right), \left(x^3 - \frac{3}{5}x\right)\right\rangle} = \sqrt{\frac{8}{175}}.$$

Therefore orthonormal polynomials are $p_0(x) = \frac{1}{\sqrt{2}}$, $p_1(x) = \sqrt{\frac{3}{2}}x$, $p_2(x) = \sqrt{\frac{5}{2}}\left(x^2 - \frac{2}{3}\right)$ and $p_3(x) = \sqrt{\frac{175}{8}}\left(x^3 - \frac{3}{5}x\right)$ and so on.

Example 7: Standard basis $\{(1, 0, 0), (0, 1, 0), (0, 0, 1)\}$ of R^3 is orthonormal basis, since $\langle(1, 0, 0), (1, 0, 0)\rangle = 1$, and so on.

If we consider another basis $B = \{(1, 0, 0), (1, 1, 0), (1, 1, 1)\}$ of R^3; with usual definition of inner product then B is not orthogonal basis, since $\langle(1, 0, 0), (1, 1, 0)\rangle = 1$ and $\langle(1, 1, 0), (1, 1, 2)\rangle = 2$.

We want to generate an orthonormal basis by Gram-Schmidt orthogonalization process from the above basis vectors.

Let $u_1 = (1, 0, 0) \neq (0, 0, 0)$ is linearly independent in R^3. To construct second vector, we choose α such that $u_2 = (1, 1, 0) + \alpha(1, 0, 0)$ is orthogonal to u_1

i.e. $\langle u_2, u_1 \rangle = \langle(1, 1, 0), (1, 0, 0) + \alpha(1, 0, 0), (1, 0, 0)\rangle = 0$

$1 + \alpha = 0 \Rightarrow \alpha = -1$.

Thus $u_2 = (1, 1, 0) - (1, 0, 0) = (0, 1, 0)$.

Now let third vector be $u_3 = (1, 1, 1) + \beta_1(1, 0, 0) + \beta_2(0, 1, 0)$

since $\langle u_3, u_1 \rangle = 0 = 1 + \beta_1 \Rightarrow \beta_1 = -1$ and $\langle u_3, u_2 \rangle = 0 = 1 + \beta_2 \Rightarrow \beta_2 = -1$.

Thus $u_3 = (1, 1, 1) - (1, 0, 0) - (0, 1, 0) = (0, 0, 1)$.

Therefore, orthogonal set, thus obtained, is $\{(1, 0, 0), (0, 1, 0), (0, 0, 1)\}$, which is same as standard basis.

Example 8: Generate a orthonormal set of polynomials from linearly independent set $\{1, x, x^2, x^3\}$ of P_3 with inner product defined as:

$$\langle p(x), q(x) \rangle = \int_{-1}^{1} \frac{1}{\sqrt{1-x^2}} p(x)q(x)\,dx, \text{ where } \frac{1}{\sqrt{1-x^2}} \text{ is weight function on the interval}$$

$(-1, 1)$.

Inner Product

Choose $p_0(x) = 1$, since 1 is non-zero vector.
Let $p_1(x) = x + \alpha$, choose α such that,

$$\langle p_0(x), p_1(x) \rangle = \int_{-1}^{1} \frac{1}{\sqrt{1-x^2}} p_0(x) p_1(x) dx = 0$$

$$\int_{-1}^{1} \frac{x + \alpha}{\sqrt{1-x^2}} dx = 0 + \alpha\pi = 0 \Rightarrow \alpha = 0, \text{ since } \frac{x}{\sqrt{1-x^2}} \text{ is an odd function.}$$

Thus $p_1(x) = x$.

Now let $p_2(x) = x^2 + \beta_1 + \beta_2 x$, such that $\langle p_2(x), p_0(x) \rangle = \langle p_2(x), p_1(x) \rangle = 0$.

i.e. $\beta_1 = -\frac{\langle x^2, 1 \rangle}{\langle 1, 1 \rangle}$, where $\langle x^2, 1 \rangle = \int_{-1}^{1} \frac{x^2}{\sqrt{1-x^2}} dx$, by taking $x = \sin\theta$

$$= 2\int_{0}^{\frac{\pi}{2}} \sin^2\theta \, d\theta = \left[x - \frac{\sin 2x}{2} \right]_{0}^{\frac{\pi}{2}} = \frac{\pi}{2}, \text{ and } \langle 1, 1 \rangle = \pi$$

therefore, $\beta_1 = -\left(\frac{\pi}{2}\right)\Big/\pi = -\frac{1}{2}$,

$$\beta_2 = -\frac{\langle x^2, x \rangle}{\langle x, x \rangle}, \quad \langle x^2, x \rangle = \int_{-1}^{1} \frac{x^3}{\sqrt{1-x^2}} dx = 0.$$

Thus $p_2(x) = x^2 - \frac{1}{2}$.

Let $p_3(x) = x^3 + \gamma_0 p_0(x) + \gamma_1 p_1(x) + \gamma_2 p_2(x)$.

$$= x^3 + \gamma_0 + \gamma_1 x + \gamma_2 \left(x^2 - \frac{1}{2} \right)$$

On taking $\langle p_3(x), p_0(x) \rangle = 0$, $\langle p_3(x), p_1(x) \rangle = 0$ and $\langle p_3(x), p_2(x) \rangle = 0$ and computing

$$\gamma_0 = -\frac{\langle x^3, 1 \rangle}{\langle 1, 1 \rangle} = 0, \quad \gamma_1 = -\frac{\langle x^3, x \rangle}{\langle x, x \rangle}, \quad \gamma_2 = \frac{\langle x^3, x^2 - \frac{1}{2} \rangle}{\langle x^2 - \frac{1}{2}, x^2 - \frac{1}{2} \rangle}.$$

Now

$$\langle x, x \rangle = \int_{-1}^{1} \frac{x^2}{\sqrt{1-x^2}} dx = \frac{\pi}{2}, \quad \langle x^3, x \rangle = \int_{-1}^{1} \frac{x^4}{\sqrt{1-x^2}} dx = 2\int_{1}^{\pi/2} \sin^4\theta \, d\theta = \frac{3\pi}{8}$$

$$\gamma_1 = -\frac{3}{4}, \quad \gamma_2 = -\frac{\langle x^3, (x^2 - \frac{1}{2}) \rangle}{\langle x^2 - \frac{1}{2}, x^2 - \frac{1}{2} \rangle}.$$

$$\langle x^3, (x^2 - \frac{1}{2}) \rangle = \int_{-1}^{1} \frac{x^3(x^2 - \frac{1}{2})}{\sqrt{1-x^2}} dx = 0, \text{ since integrand is odd}.$$

Thus $p_3(x) = x_3 - \frac{3}{4}x.$

Therefore orthogonal polynomials are $p_0(x) = 1$, $p_1(x) = x$, $p_2(x) = x^2 - \frac{1}{2}$, and $p_3(x) = x^3 - \frac{3}{4}x$.

To get orthonormal polynomials, we compute the magnitudes of polynomial vectors.

$$\|1\| = \sqrt{\langle 1, 1 \rangle} = \sqrt{\int_{-1}^{1} \frac{dx}{\sqrt{1-x^2}}} = \sqrt{\pi}, \quad \|x\| = \sqrt{\langle x, x \rangle} = \sqrt{\frac{\pi}{2}},$$

$$\left\| x^2 - \frac{1}{2} \right\| = \sqrt{\langle x^2 - \frac{1}{2}, x^2 - \frac{1}{2} \rangle} = \sqrt{\int_{-1}^{1} \frac{(x^2 - \frac{1}{2})^2}{\sqrt{1-x^2}} dx} = \sqrt{\frac{\pi}{8}}.$$

$$\left\| x^3 - \frac{3}{4}x \right\| = \sqrt{\frac{\pi}{32}},$$

since $\left\| x^3 - \frac{3}{4}x \right\|^2 = \int_{-1}^{1} \frac{(x^3 - \frac{3}{4}x)^2}{\sqrt{1-x^2}} dx$

$$= \int_{-1}^{1} \frac{1}{\sqrt{1-x^2}} \left(x^6 - \frac{3}{2}x^4 + \frac{9}{16}x^2 \right) dx, \text{ substituting } x = \sin\theta$$

$$= 2 \int_{0}^{\frac{\pi}{2}} \left(\sin^6\theta - \frac{3}{2}\sin^4\theta + \frac{9}{16}\sin^2\theta \right) d\theta$$

$$= \frac{\pi}{32}.$$

Therefore $\left\| x^3 - \frac{3}{4}x \right\| = \sqrt{\frac{\pi}{32}}$.

On dividing by the magnitudes of polynomials, we get orthonormal polynomials

$$p_0(x) = \frac{1}{\sqrt{\pi}},\ p_1(x) = \sqrt{\frac{2}{\pi}}x,\ p_2(x) = \sqrt{\frac{8}{\pi}}\left(x^2 - \frac{1}{2}\right) \text{ and } p_3(x) = \sqrt{\frac{32}{\pi}}\left(x^3 - \frac{3}{4}x\right).$$

Therefore according to the need a set of orthogonal or orthonormal vectors can be generated from a set of linearly independent vectors by Gram Schmidt orthogonalization process.

6.4 INNER PRODUCT SPACES

In this section some results are stated for completeness of the inner product.

Definition 6.4: A vector space V, with an inner product defined on it, is known as an inner product space.

Example 9: Vector space R^n with inner product for $u = (x_1, x_2, ..., x_n)$, $v = (y_1, y_2, ..., y_n)$ $\in R^n$, defined as $\langle u, v \rangle = x_1 y_1 + x_2 y_2 + , ..., + x_n y_n$ is an inner product space. It can be verified that all the conditions of inner product are satisfied.

Example 10: Consider the vector space $C[a, b]$ the space of continuous functions on the interval $[a, b]$ with inner product defined as $\langle f(x), g(x) \rangle = \int_{a}^{b} w(x) f(x) g(x) dx$, for $f(x)$, $g(x) \in C[a, b]$, where $w(x) \geq 0$ on $[a, b]$ is known as a weight function. Then $C[a, b]$ is an inner product space.

The following two inequalities are very useful for application.

Cauchy-Schwarz inequality: For all, u, v in a inner product space V,

(a) $|\langle u, v \rangle| \leq \|u\| \|v\|$, where $\|u\| = \sqrt{\langle u, u \rangle}$

From orthogonal projection, we know that

$\dfrac{\langle v, u \rangle}{\langle u, u \rangle} u$ is orthogonal projection of vector v on u.

Now magnitude of projection is always less than or equal to magnitude of v, therefore

$$\left\| \dfrac{\langle v, u \rangle}{\langle u, u \rangle} u \right\| \leq \|v\| \Rightarrow \dfrac{|\langle v, u \rangle| \|u\|}{\|u\|^2} \leq \|v\|.$$

i.e. $\quad |\langle u, v \rangle| \leq \|u\| \cdot \|v\|$

Hence Schwarz inequality.

(b) **Triangular Inequality:** for all u, v in V, $\|u + v\| \leq \|u\| + \|v\|$.

For proof of triangle inequality.

Consider $\quad \|u + v\|^2 = \langle u + v, u + v \rangle = \langle u, u + v \rangle + \langle v, u + v \rangle$

$\qquad\qquad = \langle u, u \rangle + \langle u, v \rangle + \langle v, u \rangle + \langle v, v \rangle.$

In case of real vector spaces $\langle u, v \rangle = \langle v, u \rangle$,

therefore $\|u + v\|^2 = \|u\|^2 + 2\langle u, v \rangle + \|v\|^2 \leq \|u\|^2 + 2\|u\| \cdot \|v\| + \|v\|^2$, using Schwarz inequality.

Therefore $\|u + v\|^2 \leq (\|u\| + \|v\|)^2.$

Hence $\|u + v\| \leq \|u\| + \|v\|$

Triangular inequality.

EXERCISE SET 6

1. Let $P_3(x)$ be the vector space of polynomial of degree ≤ 3, and for $p(x), q(x) \in P_3(x)$, inner product be defined $\langle f, g \rangle = \int_0^1 f(x) g(x) dx$. Compute $\langle 1 - x, x^2 + 1 \rangle$.

2. Let $u, v \in R^3$, check whether $\langle u, v \rangle = (x_1 y_1 + x_2 + y_3)$ for $u = (x_1, x_2, x_3)$, $v = (y_1, y_2, y_3)$ is an inner product?

3. Let inner product be defined by $\langle f, g \rangle = \int_{-\pi}^{\pi} f(x) g(x) dx$, compute $\langle \cos mx, \cos nx \rangle$ and $\langle \sin mx, \sin nx \rangle$ for any integers m, n.

Inner Product

4. Let $V = R^n = \{(x_1, x_2, ..., x_n) | x_i\text{'s are real numbers}\}$ be the space of n-tuples. Further Let $u = (x_1, x_2, ..., x_n) \in R^n$ and $v = (y_1, y_2, ..., y_n) \in R^n$.
   Let $\langle u, v \rangle = \max |x_i y_i|$, $i = 1, 2, ... n$ be defined in R^n, Check whether it is an inner product in R^n

5. Find a vector (x_1, x_2, x_3) which is orthogonal to the vectors $u = (1, 1, 0)$ and $v = (0, 0, 1)$ in inner product space R^3 with usual inner product defined on R^3.

6. Let $B = \{1 - x, 1 + x\}$ be in P_2, space of polynomial of degree ≤ 2.
   If $\langle p(x), q(x) \rangle = \int_{-1}^{1} p(x) q(x) dx$ be defined an inner product on P_2.
   Compute $P_2(x) = a_0 + a_1 x + a_2 x^2$, which is orthogonal to B.

7. Generate first three non-zero orthogonal polynomials on $[-1, 1]$, with respect to weight function $w(x) = x^2$, using Gram-Schmidt orthogonalization process.

8. Generate first three non-zero orthogonal polynomials on $\left[-\frac{\pi}{2}, \frac{\pi}{2}\right]$, with respect to weight function $w(x) = \cos x$, using Gram-Schmidt orthogonalization process.

9. Find the orthogonal projection of the vector $u = (1, 1, 2) \in R^3$, on the vector $v = (1, 1, 1)$, with usual inner product on R^3.

10. Show for m, n integers $\sin mx, \cos nx$ are orthogonal in space of $C[-\pi, \pi]$, with inner product defined as $\langle f, g \rangle = \int_{-\pi}^{\pi} f(x) g(x) dx$.

11. Verify traingular inequality for $\sin x$, $\cos x$ with inner product $\langle f, g \rangle = \int_{-\pi}^{\pi} f(x) g(x) dx$ in the space of $C[-\pi, \pi]$.

12. Generate first three non-zero polynomials from the Set $S = \{1, x, x^2\}$, where inner product is defined by $\langle f, g \rangle = \int_{-1}^{1} \frac{f(x) g(x)}{1 + x^2} dx$.

ANSWERS TO EXERCISE SET – 6

1. $\frac{8}{3}$, 2. No, 3. 0,0 4. Yes 5. $(1, -1, 0)$, 6. $1-x-2x^2$,

7. $p_0(x) = 1, p_1(x) = x, p_2(x) = 3 - 5x^2$ 8. $p_0(x) = 1, p_1(x) = x, p_2(x) = \left(\frac{\pi^2}{4} + 1\right) - x^2$

9. $\frac{4}{3}(1, 1, 1)$ 12. $p_0(x) = 1, p_1(x) = x, p_2(x) = (4 - \pi) - \pi x^2$.

Index

Addition of maps   4.24
Additive inverse   3.4
Adjoint matrix   5.9

Basis   3.26

Cauchy-Schwarz inequality   6.11
Column rank   4.31
Composition of maps   4.24
Consistancy of linear system   2.12

Dependence   3.13
Diagonal matrix   1.2, 2.16
Diagonalization of a matrix   5.11
Dimension   3.26

Eigenspace   5.1
Eigenvactors   5.1
Eigenvalues   5.1
Elementary row opertions   2.3
Endomorphism   1.19

Field   1.23

Gershgorin circle   5.4
Gram-schmidl-orthogonlization   6.5
Graph solution   2.2
Groups   1.10
    Semi-group   1.11
    Abelian Group   1.12

Addition modulo n 1.13
Multiplication modulo n 1.13
Cyclic   1.16
Cosets   1.17
Lagrange's Theorem   1.17
Normal   1.18

Homomorphism   1.19

Idempolent map   4.25
Identity map   4.24
Independence   3.13
Inner product   6.1
Inverse of a matrix   2.20
Inverse Transformation   4.18
Isomorphism   1.19

Kernel   1.19

Linear combination   3.13
Linear equations   2.7
Linear Transformations   4.1
Lower triangular matrix   1.3, 2.16

Matrix   1.1
    Square   1.1
    Diagonal   1.1
    Identity   1.1
    Transpose   1.5

Index

    Symmetric   1.5
    Skew–Symmetric   1.5
    Hermitian   1.6
    Skew–Hermitian   1.7
Matric of transformation   4.25
Monomorphism   1.19

Nullity   4.9
Nilpolant map   4.25
Non-linear equations   2.1
Non-singular matrix   2.16
Non-singular transform   3.18
Norm   6.1
Null space   4.7

One-one map   4.11
Orthogonal matrices   5.11
Orthogonal projection   6.5
Orthogonality   6.3
Orthonormal set   6.3
Outo map   4.11

Polynomial   3.1
Principal axes   5.16

Quadratic forms   5.15

Range space   4.7
Rank   4.9
Rank of matrix   4.31
Rank-multity theorem   4.13
Row equivalnet matrices   2.4
Row rank   4.31
Row-echelon form   2.4
Row-reduced echolen form   2.4, 2.9, 3.21

Scalars   3.2
Similar matrices   5.9
Singular Matrices   2.6
Span of a set   3.17
    Rings   1.20
Subspaces   3.9

Triangular Inequality   6.12
Trivial space   3.10

Upper Triangular matrix   2.16

Vector spaces   3.1, 3.9

Zero vector   3.2